Path Integrals For Stochastic Processes

An Introduction

Path Integrals For Stochastic Processes

An Introduction

Horacio S Wio

Instituto de Fisica de Cantabria,
Universidad de Cantabria, and CSIC, Spain

NEW JERSEY • LONDON • SINGAPORE • BEIJING • SHANGHAI • HONG KONG • TAIPEI • CHENNAI

Published by

World Scientific Publishing Co. Pte. Ltd.

5 Toh Tuck Link, Singapore 596224

USA office: 27 Warren Street, Suite 401-402, Hackensack, NJ 07601

UK office: 57 Shelton Street, Covent Garden, London WC2H 9HE

British Library Cataloguing-in-Publication Data
A catalogue record for this book is available from the British Library.

PATH INTEGRALS FOR STOCHASTIC PROCESSES
An Introduction

ISBN 978-981-4447-99-7

Printed in Singapore by World Scientific Printers.

To the memory of my parents and sister, Welka, Sofia, María Ruth; To my wife María Luz, my sons and daughter Marcelo, Mayra, Nicolás, and my grand children Sofia, Gael, Matilda, Lucas, Ilan, ..., with love.

Preface

....
caminante no hay camino
se hace camino al andar..
....
Antonio Machado

The path-integral technique has proved to be a very powerful tool in various areas of physics, both computationally and conceptually (Feynman (1948); Feynman and Hibbs (1965); Schulman (1981); Langouche *et al.* (1982); Wiegel (1986); Wio (1990); Kleinert (1990b); Khandekar and Lawande (1986); Khandekar *et al.* (2000)). It often provides an alternative route for the derivation of perturbation expansions as well as an excellent framework for non perturbative analysis. However, with few exceptions till recently, the subject of path-integrals was almost absent from standard textbooks on quantum mechanics and statistical physics (Shankar (1980); Felsager (1985); Das (1994)). As a consequence, students were missing a topic relevant for its application in field theory as well as an alternative approach to standard quantum and statistical mechanics that provides a wealth of approximation methods. However, during the last decades, many authors have tried to overcome this deficiency and have published several papers on this subject with an obvious pedagogical aim. In this way topics such as semiclassical approximations, barrier penetration, description of bound and metastable states and non exponential decay, adiabatic propagators, quantization of constrained Hamiltonians, the density matrix, and the harmonic oscillator with temperature and numerical methods among others, have been discussed within this framework (Holstein (1981a,b, 1982); Holstein and Swift (1982a,b); Mannheim (1983); Holstein (1983); MacKeown (1985); Salem and Wio (1986); Sengupta (1986); Ajanapon (1987); Larsen and Ravndal (1988); Donoghue and Holstein (1988); Gerry and Kiefer (1988); Holstein (1988a,b, 1989); Fletcher (1990); Abramson *et al.* (1991a); Cruz (1992)).

What seems to be still missing is an introductory presentation of the path-integral technique within the realm of stochastic processes (Wio (1999)). As a matter of fact, and from a historical point of view, it was in this context that path-integrals were first discussed by Wiener (Wiener (1923, 1924, 1930)), when he introduced a similar approach based on a sum over trajectories—anticipating by two decades Feynman's work on path integrals(Feynman (1948))—that were later applied by Onsager and Machlup to some Markov out of equilibrium processes in order to describe diffusion-like phenomena (Onsager and Machlup (1953a,b)).

The mathematical theory of stochastic processes has proven to be not only a useful but also a necessary tool when studying physical, chemical and biological systems under the effect of fluctuations (Haken (1978); van Kampen (2004); Risken (1983); Gardiner (2009); Mikhailov (1990); Mikhailov and Loskutov (1992); Wio (1994); Nicolis (1995); Lindenberg and Wio (2003); Wio, Deza and López (2012)). Recent theoretical and experimental studies have shown that there are even situations where fluctuations (or noise) play an essential role triggering new phenomena, solely induced by the presence of noise. A few examples of such situations are: some problems related with self-organization and dissipative structures, noise induced transitions, noise-induced-*phase*-transitions, noise sustained patterns, Brownian motors, stochastic resonance in zero-dimensional and in spatially extended systems (Horsthemke and Lefever (1984); Moss (1992); van den Broeck *et al.* (1994); Wiesenfeld and Moss (1995); Walgraef (1997); Mangioni *et al.* (1997, 2000); Jülicher *et al.* (1997); Gammaitoni *et al.* (1998); Reimann (2002); Wio, *et al.* (2002); Lindner *et al.* (2004); Sagues *et al.* (2007); Wio and Deza (2007)).

The aim of this short course is to offer a brief presentation of the path-integral approach to stochastic processes. We will focus on Markov processes, but a few aspects of non-Markov and non-Gaussian processes will also be discussed. Before starting our presentation, I would like to list a few more books, proceedings and review articles that, in addition to those indicated before, are a useful complementary reference material to this short introductory course and to the path integral approach in general (Khandekar and Lawande (1975, 1986); Papadopoulos and Devreese (1978); Marinov (1980); Brink (1985); Fox (1986); Sa-yakanit *et al.* (1989); Hänggi *et al.* (1990); Grosche and Steiner (1998); Cerdeira *et al.* (1992); Dykman and Lindenberg (1994); Wio (1999); Mazzucchi (2009)).

These notes are based on courses on path integrals taught at the Instituto Balseiro (Universidad Nacional de Cuyo) and the Physics Department (Universidad Nacional de Mar del Plata), in Argentina; at the Universitat de les Illes Balears and Universidad de Cantabria in Spain, and summer courses at San Luis, Argentina and St. Etienne de Tinée, France.

The material is organized as follows. For the sake of completeness, we start making a brief review of stochastic processes. After that, we present a derivation of the path integral representation for the propagator of Markov processes. We describe next the path expansion method as adapted to the present stochastic case. After that, we present a simple example of a space-time transformation within the path integral framework. The results of this transformation are exploited to proceed a little further on the path expansion method. We also discuss some results for non-Markov and non-Gaussian processes. Next we present a few aspects related to fractional Brownian motion. We then analyze the usefulness of the Feynman–Kac formula, and always within the stochastic framework, how to use an influence functional like procedure to eliminate irrelevant variables. Afterwards, we discuss a few more applications to different diffusive-like problems. The last chapter is devoted to comment on some aspects that have not been touched upon in these notes.

I wish to express my thanks to G. Abramson, C. Batista, C.B. Briozzo, C.E. Budde, F. Castro, P. Colet, R.R. Deza, G. Drazer, M.A. Fuentes, J. Giampaoli, P. Hänggi, G. Izús, M.N. Kuperman, K. Lindenberg, S. Mangioni, L. Pesquera, J.A. Revelli, M.A. Rodriguez, L.D. Salem, A. Sánchez, M. San Miguel, L. Schulman, U. Smilanski, D. Strier, E. Tirapegui, R. Toral, D.H. Zanette, for fruitful discussions and/or collaborations on the path-integral approach to stochastic processes. I also thank the many students who have endured with stoicism the courses on path-integrals that I have taught. Last but not least, I thank V. Grunfeld for the critical reading of an earlier version of the manuscript.

Horacio S. Wio
Santander, August 2012

Along this road
Goes no one,
This autumn eve
Matsuo Bashô

Contents

Chapter 1

Stochastic Processes: A Short Tour

1.1 Stochastic Process

We start this section by writing the evolution equation for a one dimensional dynamical system (Haken (1978); Wio (1994); Nicolis (1995); Wio (1997))

$$\frac{dx}{dt} = F(x, \zeta), \tag{1.1}$$

where x corresponds to the state variable while ζ is a control parameter. Such a parameter could be, for instance, the temperature, an external field, a reactant's controlled flux, etc, indicating the form in which the system is coupled to its surroundings. Experience tells us that it is usually impossible to keep the value of such parameters fixed, and consequently that fluctuations become relevant. Hence, the original *deterministic* equation will acquire a random or *stochastic* character.

Among the many reasons justifying the growing interest in the study of fluctuations we can point out that they present a serious impediment to accurate measurements in very sensitive experiments, demanding some very specific techniques in order to reduce their effects, and that the fluctuations might be used as an additional source of information about the system. But maybe the most important aspect is that fluctuations can produce macroscopic effects contributing to the appearance of some form of *noise-induced order* like *space-temporal patterns* or *dissipative structures* (Horsthemke and Lefever (1984); Nicolis (1995); Wio (1994); Walgraef (1997); Wio (1997); Gammaitoni *et al.* (1998); Reimann (2002); Wio, *et al.* (2002); Lindner *et al.* (2004); Sagues *et al.* (2007); Wio and Deza (2007); Wio, Deza and López (2012)).

The general character of the evolution equations of dynamical systems makes it clear why stochastic methods have become so important in different branches of physics, chemistry, biology, technology, population dynamics, economy, and sociology. In spite of the large number of different problems that arise in all these fields, there are some common principles and methods that are included in a global framework: *the theory of stochastic processes*. Here we will only briefly review the few aspects relevant for our present needs. For deeper study we refer to van Kampen (2004); Risken (1983); Horsthemke and Lefever (1984); Gardiner (2009); Wio (1994); Lindenberg and Wio (2003); Wio, Deza and López (2012).

In order to include the presence of fluctuations into our description, we write $\zeta = \zeta_0 + \xi(t)$, where ζ_0 is a constant value and $\xi(t)$ is the random or fluctuating contribution to the parameter ζ. The simplest (or lowest order) form that equation (1.1) can adopt is

$$\frac{dx}{dt} = \dot{x} = F(x, \zeta_0) + g(x, \zeta_0)\xi(t), \tag{1.2}$$

The original deterministic differential equation has been transformed into a *stochastic differential equation* (SDE), where $\xi(t)$ is called a *noise* term or stochastic process.

Any stochastic process $x(t)$ is completely specified if we know the complete hierarchy of probability densities. We write

$$P_n(x_1, t_1;\ x_2, t_2;\ ...\ ;\ x_n, t_n)\ dx_1\ dx_2\ ...dx_n, \tag{1.3}$$

for the probability that $x(t_1)$ is within the interval $(x_1, x_1 + dx_1), x(t_2)$ in $(x_2, x_2 + dx_2)$, and so on. These P_n may be defined for $n = 1, 2, ...$, and only for different times. This hierarchy fulfills some properties

i) $P_n \geq 0$

ii) P_n is invariant under permutations of pairs (x_i, t_i) and (x_j, t_j)

iii) $\int P_n\ dx_n = P_{n-1}$, and , $\int P_1\ dx_1 = 1$.

Another important quantity is the *conditional probability density* $P_{n/m}$ that corresponds to the probability of having the value x_1 at time t_1, x_2 at $t_2, \ldots, x_n$ at t_n; given that we have $x(t_{n+1}) = x_{n+1}, x(t_{n+2}) = x_{n+2}, x(t_{n+3}) = x_{n+3}, \ldots, x(t_{n+m}) = x_{n+m}$. Its definition is

$$\begin{aligned} P_{n/m}(x_1,\ t\ _1; ..; x_n, t_n \mid x_{n+1}, t_{n+1}; ..; x_{n+m}, t_{n+m}) &= \\ = \frac{P_{n+m}(x_1, t_1; \ldots; x_n, t_n; x_{n+1}, t_{n+1}; ...; x_{n+m}, t_{n+m})}{P_m(x_{n+1}, t_{n+1}; ..; x_{n+m}, t_{n+m})} \end{aligned} \tag{1.4}$$

Among the many possible classes of stochastic processes, there is one that plays a central role: *Markov Processes* (van Kampen (2004); Risken (1983); Gardiner (2009); Wio (1994); Lindenberg and Wio (2003)). For a stochastic process $x(t)$,

$$P(x_2, t_2 \mid x_1, t_1)$$

is the *conditional* or *transition* probability that $x(t_2)$ takes the value x_2, knowing that $x(t_1)$ has taken the value x_1. From this definition and (1.4) results the following identity for the *joint probability* $P_2(x_1, t_1; x_2, t_2)$ (Bayes' rule)

$$P_2(x_1, t_1; x_2, t_2) = P(x_2, t_2 \mid x_1, t_1)\, P_1(x_1, t_1). \tag{1.5}$$

A process $x(t)$ is called *Markovian* if for every set of successive times $t_1 < t_2 < .. < t_n$, the following condition holds

$$\begin{aligned} P_n(x_1, t_1, \ldots, x_n, t_n) &= P_1(x_1, t_1) P_{n-1}(x_2, t_2, \ldots, x_n, t_n \mid x_1, t_1) \\ &= P_1(x_1, t_1)\, P(x_n, t_n \mid x_{n-1}, t_{n-1}) ... P(x_2, t_2 \mid x_1, t_1), \end{aligned} \tag{1.6}$$

From this definition, it follows that a Markov process is completely determined if we know $P_1(x_1, t_1)$ and $P(x_2, t_2 \mid x_1, t_1)$. It is easy to find a relevant condition to be fulfilled for Markov processes: specifying the previous equation for the case $n = 3$ and integrating over x_2, we obtain

$$\begin{aligned} \int dx_2\, P_3(x_1, t_1, x_2, t_2, x_3, t_3) &= P_2(x_1, t_1, x_3, t_3) \\ &= P_1(x_1, t_1)\, P(x_3, t_3 \mid x_1, t_1) \\ &= \int dx_2\, P_1(x_1, t_1)\, P(x_3, t_3 \mid x_2, t_2)\, P(x_2, t_2 \mid x_1, t_1). \end{aligned} \tag{1.7}$$

For $t_1 < t_2 < t_3$ we find the identity

$$P(x_3, t_3 \mid x_1, t_1) = \int dx_2\, P(x_3, t_3 \mid x_2, t_2)\, P(x_2, t_2 \mid x_1, t_1), \tag{1.8}$$

which is the *Chapman–Kolmogorov Equation* for Markov processes. Every pair of non-negative functions $P_1(x_1, t_1)$ and $P(x_2, t_2 \mid x_1, t_1)$, adequately normalized and satisfying not only (1.8) but also

$$P_1(x_2, t_2) = \int dx_1\, P_1(x_1, t_1)\, P(x_2, t_2 \mid x_1, t_1), \tag{1.9}$$

defines a Markov process. Some typical (useful) examples of Markov processes are: the *Wiener–Levy*, the *Ornstein–Uhlenbeck* and the *Poisson* processes (van Kampen (2004); Gardiner (2009)).

Before introducing the master equation, let us briefly discuss about changing variables. Assume we have the following relation between stochastic variables y and x

$$y = f(x). \tag{1.10}$$

A familiar example could be the use of a logarithmic scale $y = \log x$. In general, the ranges of both variables will differ. The probability that y has a value between y and $y + \triangle y$ is given by

$$P(y)\triangle y \;=\; \int_{y<f(x)<y+\triangle y} dx\; P(x), \tag{1.11}$$

where the integral extends over all intervals of the range of x where the inequality is fulfilled. We can write equivalently

$$P(y) \;=\; \int dx\; \delta[f(x) - y]P(x). \tag{1.12}$$

If we have a one-to-one relation among the variables (the dimensions of x and y are the same) it is possible to invert (1.10) obtaining

$$P(y) = \mathcal{J}P(x), \tag{1.13}$$

with $\mathcal{J}$ the absolute value of the Jacobian's determinant.

1.2 Master Equation

The Chapman–Kolmogorov equation (which is only a property of the solutions for Markov processes) can be recast into a useful form. Going back to (1.8), we take $t_3 = t_2 + \delta t$ and consider the limit $\delta t \to 0$. It is clear that we have $P(x_3, t_3 \mid x_2, t_2) = \delta(x_3 - x_2)$, and it is intuitive to assume that, if $t_3 - t_2 \simeq \delta t$ (very small), the probability that a transition happens must be proportional to δt. Accordingly we adopt

$$\begin{aligned} P(x_3, t_2 + \delta t \mid x_2, t_2) \;&=\; \delta(x_3 - x_2)\; [1 - A(x_2)\; \delta t] \\ &\quad + \delta t\; W(x_3 \mid x_2) + O(\delta t^2), \end{aligned} \tag{1.14}$$

where $W(x_3 \mid x_2)$ is the *transition probability per unit time* from x_2 to x_3 (which in general could also be a function of t_2). The probability normalization tells us that

$$A(x_2) \;=\; \int W(x_3 \mid x_2)\; dx_3.$$

Substitution of the form for $P(x_3, t_2 + \delta t \mid x_2, t_2)$ into (1.8) gives

$$\begin{aligned}P(x_3, t_2 + \delta t \mid x_1, t_1) &= \int P(x_3, t_2 + \delta t \mid x_2, t_2)\, P(x_2, t_2 \mid x_1, t_1)\, dx_2 \\ &= [1 - A(x_3)\, \delta t]\, P(x_3, t_2 \mid x_1, t_1) + \delta t \int W(x_3 \mid x_2) P(x_2, t_2 \mid x_1, t_1) dx_2 \\ &= P(x_3, t_2 \mid x_1, t_1) \; - \; \delta t \int W(x_2 \mid x_3)\, P(x_3, t_2 \mid x_1, t_1)\, dx_2 \\ &\quad + \delta t \int W(x_3 \mid x_2)\, P(x_2, t_2 \mid x_1, t_1)\, dx_2. \end{aligned} \tag{1.15}$$

After rearranging and taking the limit $\delta t \to 0$ we get

$$\frac{P(x_3, t_2 + \delta t | x_1, t_1) - P(x_3, t_2 | x_1, t_1)}{\delta t} \approx \frac{\partial}{\partial t} P(x, t | x_0, t_0), \tag{1.16}$$

finally yielding

$$\frac{\partial}{\partial t} P(x, t | x_0, t_0) = \int \Big(W(x | x') P(x', t' | x_0, t_0) - W(x' | x) P(x, t | x_0, t_0) \Big) dx', \tag{1.17}$$

which corresponds to the *Master Equation* (van Kampen (2004); Gardiner (2009); Wio (1994); Lindenberg and Wio (2003); Wio, Deza and López (2012)).

The master equation is a differential form of the Chapman–Kolmogorov equation. It is an equation for the transition probability $P(x, t \mid x_0, t_0)$, and more adequate for mathematical manipulations than the Chapman–Kolmogorov equation, and it has a direct physical interpretation as a balance equation. At the same time, $W(x \mid x')\delta t$ is the transition probability during a very short time (δt). It could be evaluated by approximate methods, for instance by time dependent perturbation theory (i.e. : the *Fermi golden rule*), (van Kampen (2004); Gardiner (2009); Wio (1994)).

1.3 Langevin Equation

Brownian motion is the oldest and best known physical example of a Markov process. This phenomenon corresponds to the motion of a heavy test particle, immersed in a fluid composed of light particles in random motion. Due to the (random) collisions of the light particles against the test particle, the velocity of the latter varies in a (wide) sequence of small, uncorrelated jumps. However, similar ideas can (and have) been applied to a large variety of systems (van Kampen (2004); Gardiner (2009); Brink (1985); Wio

(1994); Lindenberg and Wio (2003); Wio, Deza and López (2012)). To simplify the presentation we restrict the description to a one dimensional system.

We will give a simple quantitative picture of Brownian motion. We start by writing Newton's equation as

$$m\,\dot{v} \;=\; F(t) \;+\; f(t), \tag{1.18}$$

where m is the mass of the Brownian particle, v its velocity, $F(t)$ the force due to some external field (i.e. gravitational, electrical for charged particles, etc), and $f(t)$ is the force produced by the collisions of fluid particles against the test particle. Due to the rapid fluctuations in v, we have two effects. On one hand a *systematic* one, i.e., a kind of *friction* that tends to slow down the particle, while on the other hand, a *random* contribution from the random hits of the fluid particle. If the mass of the test particle is much larger than the mass of the fluid particles (implying that the fluid *relaxes* faster than the test particle, allowing us to assume that it is always in equilibrium), we can write

$$\frac{1}{m}f(t) \;=\; -\;\gamma\,v \;+\; \xi(t). \tag{1.19}$$

In the r.h.s., γ is the friction coefficient, and the minus sign in the first term indicates that this contribution should oppose the motion (as a well behaved friction term). The second term corresponds to the stochastic or random contribution, since we have separated the systematic contribution in the first term, and this random contribution averages to zero : $\langle\xi(t)\rangle \;=\; 0$ (where the average is over an *ensemble* of noninteracting Brownian particles). In order to define the so called *Langevin force* (or *white noise*) it is required that

$$\langle\xi(t)\xi(t')\rangle \;=\; D\;\delta(t-t'). \tag{1.20}$$

We will not consider higher order moments, but it is clear that to fully characterize the fluctuating force, we need the whole hierarchy of moments (van Kampen (2004); Gardiner (2009)).

With the above indicated arguments, and without an external field, (1.18) adopts the form

$$\dot{v} \;=\; -\;\gamma\,v \;+\; \xi(t), \tag{1.21}$$

which is known as the *Langevin equation.* This is the simplest example of a SDE (that is, a differential equation whose coefficients are random functions with known stochastic properties). Hence $v(t)$ is a stochastic process, with a given initial condition. For details we refer the reader to van Kampen (2004); Risken (1983); Horsthemke and Lefever (1984); Gardiner (2009); Wio (1994); Lindenberg and Wio (2003); Wio, Deza and López (2012).

When an external field is present, we have the pair of equations

$$\begin{aligned} \dot{x} &= v \\ \dot{v} &= \frac{1}{m} F(x) - \gamma\, v + \xi(t). \end{aligned} \tag{1.22}$$

After differentiating the first one and substituting in the second, it adopts the form

$$\ddot{x} = \frac{1}{m} F(x) - \gamma\, \dot{x} + \xi(t). \tag{1.23}$$

In the case of large friction (γ very large), through an *adiabatic elimination* ($\ddot{x} \simeq 0$) (Haken (1978); van Kampen (2004); Gardiner (2009); Wio (1994); Wio, Deza and López (2012)), we can rewrite the last equation as

$$\dot{x} = - \frac{\partial}{\partial x} V(x) + \xi(t), \tag{1.24}$$

where $\frac{\partial}{\partial x} V(x) = - F(x)$, and m and γ have been absorbed into the different terms. The last result corresponds to the problem of *diffusion in a field* in the *overdamped* case (van Kampen (2004); Risken (1983); Gardiner (2009); Wio (1994); Lindenberg and Wio (2003)).

1.4 Fokker–Planck Equation

Let us now return to the Master Equation (1.17). We assume that x is a continuous variable, and that its changes correspond to *small jumps* (or variations). In this case it is possible to derive, starting from the Master Equation, a differential equation. The transition probability $W(x \mid x')$ will decay very fast as a function of $|x - x'|$. We could then write $W(x \mid x') = W(x', \xi)$, where $\xi = x - x'$ corresponds to the size of the jump. The Master Equation will take the form

$$\begin{aligned} \frac{\partial}{\partial t} P(x, t \mid x_0, t_0) = &\int W(x - \xi, \xi) P(x - \xi, t \mid x_0, t_0) d\xi \\ &- P(x, t \mid x_0, t_0) \int W(x, -\xi) d\xi. \end{aligned} \tag{1.25}$$

Following our assumption of small jumps, and the additional argument that P varies slowly with x, we make a Taylor expansion in ξ that gives

$$\frac{\partial}{\partial t}P(x,t \mid x_0,t_0) = \int \Big[W(x,\xi)\ P(x,t \mid x_0,t_0) - \xi\frac{\partial}{\partial x}W(x,\xi)\ P(x,t \mid x_0,t_0)$$
$$+ \xi^2\ \frac{\partial^2}{\partial x^2}W(x,\xi)\ P(x,t \mid x_0,t_0) - ...\Big]\ d\xi$$
$$- P(x,t \mid x_0,t_0)\int W(x,-\xi)d\xi. \qquad (1.26)$$

As the first and the last terms are equal (in the latter changing $-\xi$ by ξ, as well as the integration limits), we get

$$\frac{\partial}{\partial t}\ P(x,t \mid x_0,t_0) \ = \ \sum_{\nu=1}^{\infty}\frac{(-1)^{\nu}}{\nu!}\ \frac{\partial^{\nu}}{\partial x^{\nu}}\ \alpha_{\nu}(x)\ P(x,t \mid x_0,t_0), \quad (1.27)$$

with $\alpha_{\nu}(x) = \int \xi^{\nu}\ W(x,\xi)d\xi$. This result corresponds to the *Kramers–Moyal expansion* of the Master Equation (van Kampen (2004); Risken (1983); Gardiner (2009); Wio (1994); Wio, Deza and López (2012)). Up to this point we have gained nothing. However, there could be situations where, for $\nu > 2$, the α_{ν} are either zero or very small (even though there are no *a priori* criteria about the relative size of the terms). If this is the case, we have

$$\frac{\partial}{\partial t}P(x,t \mid x_0,t_0) = -\ \frac{\partial}{\partial x}\alpha_1(x)\ P(x,t \mid x_0,t_0)$$
$$+\frac{1}{2}\ \frac{\partial^2}{\partial x^2}\ \alpha_2(x)\ P(x,t \mid x_0,t_0), \qquad (1.28)$$

which corresponds to the *Fokker–Planck equation* (van Kampen (2004); Risken (1983); Gardiner (2009); Wio (1994); Lindenberg and Wio (2003); Wio, Deza and López (2012)).

Let us look at a couple of examples. For the Wiener–Levy process we find that $\alpha_1(x) = 0$ and $\alpha_2(x) = 1$, then

$$\frac{\partial}{\partial t}\ P(x,t \mid x_0,t_0) \ = \ \frac{\partial^2}{\partial x^2}\ P(x,t \mid x_0,t_0),$$

while for the case of the Ornstein–Uhlenbeck process we find $\alpha_1(x) = x$, $\alpha_2(x) = D$, and we get

$$\frac{\partial}{\partial t}\ P(x,t \mid x_0,t_0) \ = \ -\ \frac{\partial}{\partial x}\ x\ P(x,t \mid x_0,t_0)\ +\ \frac{\partial^2}{\partial x^2}\ P(x,t \mid x_0,t_0).$$

Equation (1.28) corresponds to a *nonlinear* Fokker–Planck equation (due to the dependence of $\alpha_1(x)$ and $\alpha_2(x)$ on x), which is the result of poorly grounded assumptions (i.e., the criteria to decide where to cut the expansion, etc). Even worse, it is **not a systematic** approximation to the Master Equation. However, there is a procedure due to van Kampen that does make it possible to build up such a systematic procedure, but we will not discuss it here and instead refer the reader to van Kampen (2004); Gardiner (2009); Wio (1994); Wio, Deza and López (2012).

Consider the long time limit, where we expect that the system will reach a stationary behavior (that is: $\partial_t P(x, t \mid x_0, t_0) = 0$). In such a case we will have that

$$0 = -\frac{d}{dx}\alpha_1(x)\, P_{st}(x) + \frac{D}{2}\frac{d^2}{dx^2}\, P_{st}(x), \tag{1.29}$$

where in order to simplify we have taken $\alpha_2(x) =$ const.$= D$. The stationary distribution turns out to be

$$P_{st}(x) \simeq \mathcal{N} e^{-\int dx' \alpha_1(x')/D}. \tag{1.30}$$

The exponent in the last equation allows us to define the (*non equilibrium*) *potential* $U(x)$ through

$$U(x) = -\int dx' \alpha_1(x'). \tag{1.31}$$

1.5 Relation Between Langevin and Fokker–Planck Equations

Here we give a brief and more or less formal (but not completely rigorous from a mathematical point of view) presentation of the relation between *stochastic differential equations* (SDE) of the *Langevin type* (LE), and *Fokker–Planck equations* (FPE). We start considering a general form for the one-dimensional SDE as indicated in (1.2):

$$\dot{x}(t) = \frac{dx(t)}{dt} = f(x(t), t) + g(x(t), t)\, \xi(t) \tag{1.32}$$

where $\xi(t)$ is a *white noise* with

$$\langle\, \xi(t)\, \rangle = 0 \quad \text{and} \quad \langle\, \xi(t)\, \xi(t')\, \rangle = \delta(t - t')$$

as in (1.19) and (1.20), with $D = 1$. We made the usual assumption that the process is Gaussian. However, $\xi(t)$ is not a well defined stochastic process. In a loose way, it could be considered as the derivative of the well defined *Wiener process*, but such a derivative does not really exist (van Kampen (2004); Gardiner (2009)). We now integrate (1.32) over a short time interval δt

$$x(t+\delta t) \; - \; x(t) \; = \; f(x(t),t)\,\delta t \; + \; g(x(t),t)\,\xi(t)\,\delta t. \qquad (1.33)$$

It is worth commenting here that in writing the above equation we have adopted the so called *Ito* prescription. The two most usual prescriptions are the Ito and the Stratonovich ones (Gardiner (2009)). The first corresponds to evaluating $g(x(t),t)\,\xi(t)$ at the beginning of the interval $[t, t+\delta t]$, while the second corresponds to evaluating it in the middle of that interval (i.e. $t^* = t + \delta t/2$). This is a point we will briefly comment in the next chapter.

As $x(t)$ is a Markov process, it is well defined if we are able to determine its probability distribution $P_1(x,t)$ as well as its conditional probability distribution $P(x,t \mid x',t')(t \; > \; t')$. In order to obtain an equation for the latter quantity, we define now a *conditional average*, corresponding to the average of a function of the stochastic variable x (say $F(x)$), given that x has the value y at $t' \; < \; t$:

$$\langle F(x(t)) \mid x(t') = y \rangle \; = \; \ll F(x(t)) \gg \; = \; \int dx' F(x') \; P(x',t \mid y,t'). \qquad (1.34)$$

Due to the property $P(x,t \mid x',t) \; = \; \delta(x - x')$, we have

$$\langle F(x(t)) \mid x(t) = y \rangle \; = \; F(y). \qquad (1.35)$$

We use now this definition in order to obtain the first few *conditional moments* of $x(t)$.

$$\begin{aligned} \ll \Delta x(t) \gg = \langle x(t+\delta t) \mid x(t) = x \rangle \; = \\ = \ll \; f(x(t),t)\,\delta t \; \gg \; + \; \ll g(x(t),t)\,\xi(t)\,\delta t \; \gg . \end{aligned} \qquad (1.36)$$

The result shown in (1.35) indicates that $\ll f(x(t),t)\,\delta t \gg \; = f(x(t),t)\,\delta t$, and also that $\ll g(x(t),t)\,\xi(t)\,\delta t \gg = g(x(t),t) \; \ll \xi(t) \gg \delta t = 0$, resulting in

$$\ll \Delta x(t) \gg \; = \; f(x(t),t)\,\delta t. \qquad (1.37)$$

For the second moment we need to resort to properties of the Wiener process; i.e. using that $\xi(t)\,\delta t \; = \; \int_t^{t+\delta t} dt \xi(t') \; = \; d\Delta W(t)$, where $W(t)$ is the Wiener process, and $\langle [\xi(t)\,\delta t]^2 \rangle \; \simeq \; \langle \Delta W(t)^2 \rangle \; = \; \Delta t$; to obtain

$$\ll \Delta x(t)^2 \gg \; = \; g(x(t),t)^2 \delta t \; + \; O(\delta t^2). \qquad (1.38)$$

Let us now consider an arbitrary function $R(x)$, and evaluate its conditional average. Using the Chapman–Kolmogorov equation

$$\int dx\ R(x)\ P(x,t+\delta t\mid y,s) = \int dz\ P(z,t\mid y,s) \int dx\ R(x)\ P(x,t+\delta t\mid z,t), \quad (1.39)$$

We can expand $R(x)$ in a Taylor series around z, since for $\delta t \simeq 0$ we know that $P(x,t+\delta t\mid z,t) \simeq \delta(x-z)$, and that only a neighborhood of z will be relevant. If we also remember the normalization condition for $P(z,t\mid y,s)$, integrate by parts and use (1.37) and (1.38) we obtain an equation that, after arranging terms and taking the limit $\delta t \to 0$, gives

$$\int dx\ R(x) \left(\frac{\partial}{\partial t}\ P(x,t\mid y,s) - (-\frac{\partial}{\partial x}\ [f(x,t)\ P(x,t\mid y,s)] + \frac{1}{2}\frac{\partial^2}{\partial x^2}[g(x,t)^2\ P(x,t\mid y,s)])\right) = 0 \quad (1.40)$$

Due to the arbitrariness of the function $R(x)$, we arrive at the condition

$$\frac{\partial}{\partial t}\ P(x,t\mid y,s) = -\frac{\partial}{\partial x}\Big[f(x,t)\ P(x,t\mid y,s)\Big] + \frac{1}{2}\frac{\partial^2}{\partial x^2}\Big[g(x,t)^2\ P(x,t\mid y,s)\Big], \quad (1.41)$$

which is the desired Fokker–Planck equation for $P(x,t\mid y,s)$, the transition probability associated with the stochastic process driven by the SDE (1.32).

We interrupted this brief overview on stochastic processes at this point, and start to discuss their path integral representation.

Chapter 2

The Path Integral for a Markov Stochastic Process

2.1 The Wiener Integral

Here we focus on one-dimensional Markovian processes describable through Langevin or Fokker–Planck equations as discussed in the previous chapter. The most general form of the Langevin equation we will consider is (see (1.32))

$$\dot{q} = f(q,t) + g(q,t)\xi(t) \tag{2.1}$$

where $f(q,t)$ and $g(q,t)$ are known (smooth) functions, and $\xi(t)$ is a Gaussian white noise with zero mean and δ-correlated. The form of the FPE, which could be related with such a Langevin equation is (in the Ito prescription) given by (1.41). As has been discussed in the last section, $P(q,t \mid q',t')$ fulfills the Chapman–Kolmogorov equation ($t_1 < t_2 < t_3$)

$$P(q_3,t_3 \mid q_1,t_1) \;=\; \int_{-\infty}^{\infty} dq_2 P(q_3,t_3 \mid q_2,t_2) P(q_2,t_2 \mid q_1,t_1). \tag{2.2}$$

Such an equation allows, by making a partition of the time interval in N steps: $t_0 < t_1 < \ldots. < t_f$, with $t_j = t_o + j\,(t_f - t_o)/N$, to obtain a path-dependent representation of the propagator. With the given partition, we reiterate (2.2) n-times and write

$$P(q_f,t_f \mid q_0,t_0) = \int_{-\infty}^{\infty} \cdots \int_{-\infty}^{\infty} dq_1 dq_2 \ldots dq_{N-1}\, P(q_f,t_f \mid q_{N-1},t_{N-1}) \ldots$$

$$\ldots P(q_2,t_2 \mid q_1,t_1)\; P(q_1,t_1 \mid q_o,t_o). \tag{2.3}$$

The last expression can be interpreted as an integration over all possible paths that the process could follow (corresponding to the different values of the sequence: $q_0, q_1, q_2, \ldots.., q_N = q_f$). In the last expression we need the form of the *short time propagator* $P(q_{j+1}, t_{j+1} \mid q_j, t_j)$ in order to find the more conventional representation of the integration over paths. However we follow an alternative way.

Now, the probability that at a given time t, the process takes a value between a and b is given by

$$\int_a^b dq\, P(q, t \mid q_0, t_0) \tag{2.4}$$

Likewise, the probability that the process, starting at $q = q_0$ at $t = t_0$, has a value between a_1 and b_1 at t_1, between a_2 and b_2 at $t_2, \ldots$, between a_{N-1} and b_{N-1} at t_{N-1}(with $a_j < b_j$ and $t_j < t_{j+1}$), and reaching q_N at t_N, will be given by

$$\int_{a_1}^{b_1} \int_{a_2}^{b_2} \cdots \int_{a_{N-1}}^{b_{N-1}} dq_1 dq_2 \ldots dq_{N-1} P(q_N, t_N \mid q_{N-1}, t_{N-1}) \ldots \\ \ldots P(q_2, t_2 \mid q_1, t_1) P(q_1, t_1 \mid q_0, t_0). \tag{2.5}$$

If we increase the number of time slices within the time partition where the intervals (a_j, b_j) are specified, and at the same time take the limit $b_j - a_j \to 0$, the trajectory is defined with higher and higher precision. Clearly, a requisite is that the trajectories be continuous. This happens in particular for the Wiener process (van Kampen (2004); Gardiner (2009)). With all this in mind (2.3) can be interpreted as an integration over all the possible paths that the process could follow (corresponding to the different values of the sequence $q_o, q_1, q_2, \ldots, q_{N-1}, q_f$). For the Wiener process we have that

$$P(W_2, t_2 \mid W_1, t_1) = \frac{1}{\sqrt{2\pi D\, (t_2 - t_1)}} \exp\left[-\frac{1}{2D(t_2 - t_1)} [W_2 - W_1]^2 \right]. \tag{2.6}$$

For $N \to \infty$, we can define a *measure* in the path-space known as the *Wiener measure* (Schulman (1981); Wio (1990)). By substituting (2.6) into (2.5) we get

$$\prod_{j=1}^{N} \left(\frac{dW_j}{\sqrt{4\pi\epsilon D}} \right) \exp\left[-\frac{1}{4D\epsilon} \sum_j (W_j - W_{j-1})^2 \right], \tag{2.7}$$

which is the desired probability of following a given path. Figure 2.1 depicts a "practical" example of such a trajectory, while Fig. 2.2 indicates alternative trajectories.

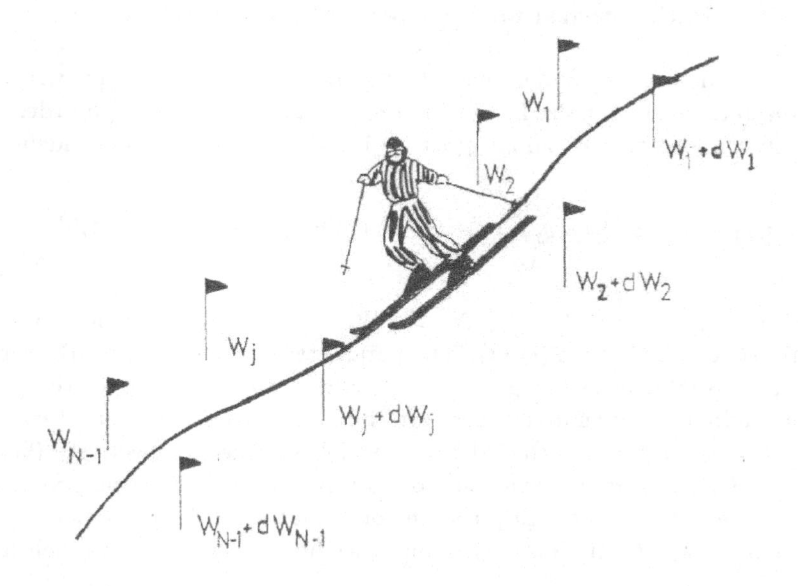

Fig. 2.1 A *practical* realization of a typical trajectory.

When $\epsilon \to 0$ and $N \to \infty$, we can write the exponential in (2.7) in the continuous limit as

$$e^{-\frac{1}{4D}\int_{t_o}^{t} d\tau\left(\frac{dW}{d\tau}\right)^2}. \tag{2.8}$$

If we integrate the expression in (2.5) over all the intermediate points (which is equivalent to a sum over all the possible paths), as all the integrands are Gaussian, and the convolution of two Gaussian is again a Gaussian, we recover the result of (2.8) for the probability density of the Wiener process. Hence, we have expressed the probability density as a **Wiener** path integral (Schulman (1981); Langouche *et al.* (1982))

$$P(W,t \mid W_0,t_0) = \int \mathcal{D}[W(\tau)]\; e^{-\frac{1}{4D}\int_{t_o}^{t} d\tau\left(\frac{dW}{d\tau}\right)^2}, \tag{2.9}$$

where the expression inside the integral represents the continuous version of the integral of (2.5), over all possible values of the intermediate points $\{W_j\}$.

2.2 The Path Integral for a General Markov Process

Let us go now back to the general SDE in (2.1). We start by writing the discrete version of the Langevin equation given by (1.33) (in order to simplify the notation we adopt $g(q,t) = 1$ and $f(q,t)$ to be independent of t)

$$q_{j+1} - q_j \simeq \{\alpha f(q_{j+1}) + (1-\alpha) f(q_j)\} \epsilon + [W_{j+1} - W_j], \tag{2.10}$$

where $\epsilon = t_{j+1} - t_j = (t_f - t_o)/N$, and $W_j = W(t_j)$ is the Wiener process (just formally, $dW(t) \simeq \xi(t)\, dt$). The parameter $\alpha (0 \leq \alpha \leq 1)$ is arbitrary, the most usual choices being $\alpha = 0$, $1/2$ and 1, corresponding to the pre-point, midpoint and endpoint discretization, respectively. The first two are related to the so called Ito and Stratonovich schemes, respectively (Gardiner (2009)). Clearly, for the case of additive noise, there is no problem. However, for the case of multiplicative noise, that is when we have a general function for $g(q,t)$, the discretization procedure becomes a more delicate issue.

In addition to the above indicated discretization scheme, the are many other that yield the same continuous limit. For instance, instead of the form adopted before

$$\int_{t_j}^{t_{j+1}=t_j+\epsilon} f(q)ds \approx \{\alpha f(q_{j+1}) + (1-\alpha) f(q_j)\} \epsilon$$

we can also use

$$\int_{t_j}^{t_{j+1}=t_j+\epsilon} f(q)ds \approx f(\alpha q_{j+1} + (1-\alpha) q_j)\ \epsilon.$$

The last is the one to be used when discussing the recovering of the Fokker–Planck equation.

We don't want to deep into this aspect here, but just call the attention, and alert the reader about the existence of several possible discretization procedures for the path integral representation of the transition probability. And this is valid in both, configuration and phase space representations. A detailed discussion on the different discretization procedures, the equivalence and/or relations among them, as well as its relation in the quantum case to operator ordering, could be found in Langouche *et al.* (1982). We will not go into the usual difficulties related to this problem, but will keep

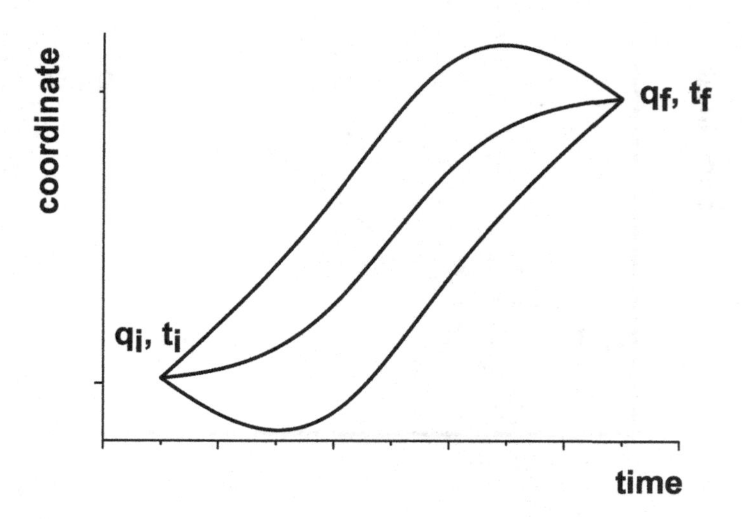

Fig. 2.2 A few possible trajectories starting at (q_i, t_i) and reaching (q_f, t_f).

this parameter in order to show the dependence of the final *Lagrangian* on it.

According to the previous results, the probability that

$$W(t_0) = 0; W_1 < W(t_1) < W_1 + dW_1; \ldots .; W_N < W(t_N) < W_N + dW_N$$

is given by

$$P(\{W_j\})\,\Delta\{W_j\} = \prod_{j=1}^{N}\Big(\frac{dW_j}{\sqrt{4\pi\epsilon D}}\Big)\ \exp\left[-\,\frac{1}{4D\epsilon}\,\sum_j (W_j\,-\,W_{j-1})^2\right]. \tag{2.11}$$

As our interest is to have the corresponding probability in q-space, we need to transform the probability given in the last equation. As is well known, to do this we need the Jacobian of the transformation connecting both sets of stochastic variables ($\{W_j\} \,\to\, \{q_j\}$). To find it we write (2.10) as

$$W_j \;=\; q_j \;-\; q_{j-1} \;-\; \{\alpha\, f(q_j) \;+\; (1-\alpha)\, f(q_{j-1})\}\,\epsilon \;+\; W_{j-1}. \tag{2.12}$$

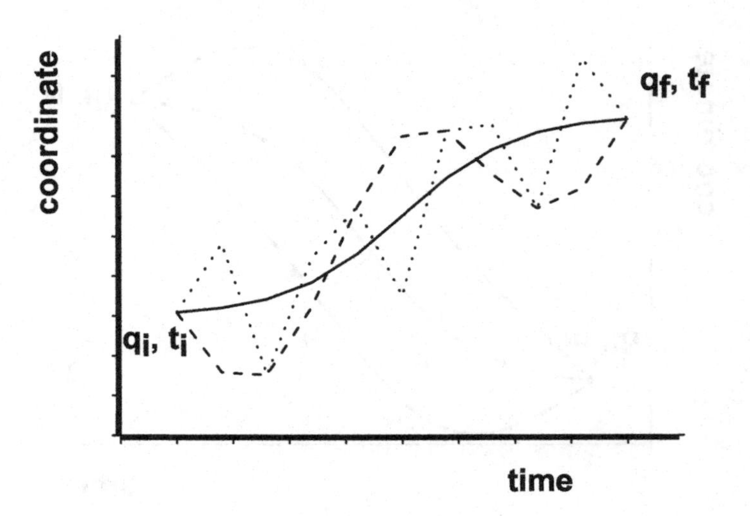

Fig. 2.3 Discrete versions of a continuous trajectory.

The Jacobian is given by

$$\mathcal{J} = \det\left(\frac{\partial W_j}{\partial q_k}\right) = \prod_{j=1}^{N}\left(1 - \epsilon\,\alpha\,\frac{df(q_j)}{dq_j}\right). \tag{2.13}$$

For $\epsilon \to 0$, it can be approximated as

$$\mathcal{J} = e^{-\epsilon\,\alpha\sum \frac{df(q_j)}{dq_j}}. \tag{2.14}$$

Now, remembering that $P(\{q_j\}) = \mathcal{J}\, P(\{W_j\})$, and taking into account that the conditional probability $P(q, t \mid q_0, t_0)$ is given as a sum over all the possible paths, we get

$$\begin{aligned} P(q, t \mid q_0, t_0) = & \int_{-\infty}^{\infty} \dots \int_{-\infty}^{\infty} \left(\frac{1}{4\pi\epsilon D}\right)^{N/2} dW_1 dW_2 \dots dW_N\, \delta(q_f - q_N) \\ & \times \exp\left(-\frac{1}{4D\epsilon}\sum_j (W_j - W_{j-1})^2\right). \end{aligned} \tag{2.15}$$

Replacing (2.10) into (2.15) and going to the continuous limit, the different terms in the exponentials yield

$$\lim_{N\to\infty,\epsilon\to0} \epsilon\,\alpha \sum_j \frac{df(q_j)}{dq_j} \to \alpha \int_{t_0}^{t} ds\, \frac{df(q(s))}{dq}$$

$$\lim_{N\to\infty,\epsilon\to0} \frac{\epsilon}{2} \sum_j \left(\alpha\, f(q_{j+1}) \;+\; (1-\alpha)\, f(q_j)\right)^2 \to \frac{1}{2} \int_{t_0}^{t} ds\, f(q(s))^2$$

$$\lim_{N\to\infty,\epsilon\to0} \frac{\epsilon}{2} \sum_j \left(\frac{(q_{j+1}-q_j)}{\epsilon}\right)^2 \to \frac{1}{2} \int_{t_0}^{t} ds\, \dot{q}(s)^2$$

$$\lim_{N\to\infty,\epsilon\to0} \sum_j (q_{j+1}-q_j)\,(\alpha f(q_{j+1}) + (1-\alpha) f(q_j)) \to \int_{t_0}^{t} dq f(q(s)). \tag{2.16}$$

Hence, we can write

$$P(q,t \mid q_0,t_0) \;=\; \int \mathcal{D}[q(t)]\; e^{-\frac{1}{D} \int_{t_0}^{t} ds\; \mathcal{L}[q(s),\dot{q}(s)]}, \tag{2.17}$$

where

$$\mathcal{L}[q(s),\dot{q}(s)] \;=\; \frac{1}{4}\left(\dot{q}(s) \;-\; f(q(s))\right)^2 \;+\; \alpha\, \frac{df(q(s))}{dq}, \tag{2.18}$$

is the *stochastic Lagrangian*, also called Onsager–Machlup functional (Onsager and Machlup (1953a,b); Langouche *et al.* (1982)). The dependence of the Lagrangian on α is clearly seen in the last expression. In (2.17), $\mathcal{D}[q]$ corresponds to the differential of the path, that is, the continuous expression of the discrete product of differentials in (2.15). We can also identify the *stochastic action* through

$$\mathcal{S}[q(t)] \;=\; \int_{t_0}^{t} ds\; \mathcal{L}[q(s),\dot{q}(s)], \tag{2.19}$$

that allows to write Eq.(2.17) as

$$P(q,t \mid q_0,t_0) \;=\; \int \mathcal{D}[q(t)]\; e^{-\,\frac{1}{D}\mathcal{S}[q(t)]}. \tag{2.20}$$

This last expression is the one we will exploit in the following.

2.3 The Recovering of the Fokker–Planck Equation

As in the quantum mechanics case, it is interesting to reassure oneself that the indicated form of the propagator allows us to recover the original Fokker–Planck equation. To verify this point, and according to the previous results, we can write a discrete version of the propagator (or conditional probability) as

$$\begin{aligned}
P(q, t+\epsilon \mid q_0, t_0) &= \int_{-\infty}^{\infty} \cdots \int_{-\infty}^{\infty} \left(\frac{1}{4\pi\epsilon D}\right)^{\frac{N+2}{2}} dq_1 dq_2 \ldots dq_{N+1} \\
&\quad \times \exp\left(-\frac{\epsilon}{4D} \sum_{j}^{N+2} \left(\frac{q_j - q_{j-1}}{\epsilon} - f(\frac{q_j + q_{j-1}}{2}) \right)^2 \right) \\
&= \int_{-\infty}^{\infty} \left(\frac{1}{4\pi\epsilon D}\right)^{\frac{1}{2}} \exp\left(-\frac{\epsilon}{4D} \left(\frac{q - q_{N+1}}{\epsilon} - f(\frac{q + q_{N+1}}{2}) \right)^2 \right) \\
&\quad \times P(q_{N+1}, t \mid q_0, t_0), \qquad (2.21)
\end{aligned}$$

where in the last line we have extracted the final integration point, and identified the rest with $P(q_{N+1}, t \mid q_0, t_0)$. Note that the discretization form we have adopted here, as the noise is additive, results equivalent to the one we used previously. See also the brief comments about the discretization procedures after (2.10).

Now we define the following change of variables $\eta = q - q_{N+1}$ that could be written as $q_{N+1} = q - \eta$, and $d\eta = -dq_{N+1}$. We also have

$$\frac{q + q_{N+1}}{2} = \frac{q + q - \eta}{2} = q - \frac{\eta}{2}.$$

Hence, the previous equation can be written as

$$\begin{aligned}
P(q, t+\epsilon \mid q_0, t_0) &= \int_{-\infty}^{\infty} d\eta \left(\frac{1}{4\pi\epsilon D}\right)^{1/2} \\
&\quad \times \exp\left(- \frac{\epsilon}{4D} \left(\frac{\eta}{\epsilon} - f(q - \frac{\eta}{2}) \right)^2 \right) P(q - \eta, t \mid q_0, t_0) \\
&\approx \int_{-\infty}^{\infty} d\eta \left(\frac{1}{4\pi\epsilon D}\right)^{1/2} \\
&\quad \times \exp\left(- \frac{\epsilon}{4D} \left(\left(\frac{\eta}{\epsilon}\right)^2 - 2\frac{\eta}{\epsilon} f(q - \frac{\eta}{2}) + f(q - \frac{\eta}{2})^2 \right) \right) \\
&\quad \times P(q - \eta, t \mid q_0, t_0), \qquad (2.22)
\end{aligned}$$

that corresponds to a Gaussian integration in η. As ϵ is very small, only small values of η will contribute to the integral. Hence, we can expand $f(q - \frac{\eta}{2}))$ and $P(q - \eta, t \mid q_0, t_0)$ around $\eta \approx 0$.

We have

$$P(q - \eta, t \mid q_0, t_0) \approx P(q, t \mid q_0, t_0) - \eta \frac{\partial}{\partial q} P(q, t \mid q_0, t_0)$$
$$+ \frac{\eta^2}{2} \frac{\partial^2}{\partial q^2} P(q, t \mid q_0, t_0) + O(\eta^3).$$

Inserting this and also the expansion of $f(q - \eta)$ we get

$$P(q, t + \epsilon \mid q_0, t_0) \approx \int_{-\infty}^{\infty} \left(\frac{1}{4\pi\epsilon D} \right)^{1/2} d\eta$$
$$\times \left(1 - \frac{1}{2D} \left[\eta f(q) - \frac{\eta^2}{2} f'(q) \right] - \frac{\epsilon}{4D} f(q)^2 .... \right)$$
$$\times \left(P(q, t \mid q_0, t_0) - \eta \frac{\partial}{\partial q} P(q, t \mid q_0, t_0) + \frac{\eta^2}{2} \frac{\partial^2}{\partial q^2} P(q, t \mid q_0, t_0) + ... \right)$$

whose integration yields

$$\begin{aligned} P(q, t + \epsilon \mid q_0, t_0) &\approx P(q, t \mid q_0, t_0) + \epsilon \left[\frac{\partial}{\partial q} \left(f(q) P(q, t \mid q_0, t_0) \right) \right. \\ &\qquad .... \left. + D \frac{\partial^2}{\partial q^2} P(q, t \mid q_0, t_0) + 0(\epsilon^2) \right] . \end{aligned} \tag{2.23}$$

Now, taking the limit $\epsilon \to 0$, and identifying

$$\frac{1}{\epsilon} \left(P(q, t + \epsilon \mid q_0, t_0) - P(q, t \mid q_0, t_0) \right) \approx \frac{\partial}{\partial t} P(q, t \mid q_0, t_0),$$

we recover the original Fokker–Planck equation.

2.4 Path Integrals in Phase Space

Similarly to what is done in a quantum mechanical framework, we can also obtain, within the present context, a path integral representation for the propagator in phase space (Marinov (1980); Langouche *et al.* (1982)). In order to show how to get it, we will exploit a notation that resembles the *bra* and *ket* representation in quantum mechanics (Landau and Lifshitz (1958)).

Let us recall the Fokker–Planck equation

$$\frac{\partial}{\partial t}P(q,t \mid q_0,t_0) = -\frac{\partial}{\partial q}\left(f(x)P(q,t \mid q_0,t_0)\right) + D\frac{\partial^2}{\partial q^2}P(q,t \mid q_0,t_0)$$

$$= -LP(q,t \mid q_0,t_0), \tag{2.24}$$

and, as in Feynman and Hibbs (1965), we define the *temporal evolution* operator through

$$P(q,t \mid q_0,t_0) = \langle q,t \mid q_0,t_0\rangle = \langle q \mid U(t,t_0) \mid q_0\rangle.$$

The kinetic equation fulfilled by the operator U results

$$\frac{\partial}{\partial t}U(t,t') = -L(\hat{p},\hat{q})U(t,t'), \tag{2.25}$$

whose solution is

$$U(t,t') = \exp\left(-\, L(\hat{p},\hat{q})[t-t']\right),$$

where p will be defined latter, and with the initial condition

$$U(t,t) = 1,$$

and the condition $\langle q \mid q'\rangle = \delta(q-q')$.

As we have done before, we can now use this notation to express the propagator in terms of a set of integrations over intermediate times (or time slicing)

$$\langle q \mid U(t,t_0) \mid q_0\rangle = \int \dots \int \prod_{j=1}^{M} dq_j \prod_{j=1}^{M+1} \langle q_j \mid U(t_j,t_{j-1}) \mid q_{j-1}\rangle, \tag{2.26}$$

where $t_j = t_{j-1} + \epsilon = t_0 + j\epsilon$, so $\epsilon = \frac{t-t_0}{M+1}$, and $t_{M+1} = t$.

We now resort to the so called Trotter's formula (see for instance Schulman (1981)). For operators A and B, and a parameter λ we have that

$$\exp\left(\lambda(A+B)\right) = \exp\left(\lambda A\right)\exp\left(\lambda B\right)\exp\left(\frac{\lambda^2}{2}[A,B] + O(\lambda^3)\right),$$

where $[A,B]$ indicates the commutator of operators A and B. For $\lambda \ll 1$ the above expression reduces to

$$\approx \exp\left(\lambda A\right)\exp\left(\lambda B\right) + O(\lambda^2).$$

Hence, for very small ϵ (or M large enough) we have

$$\langle q_j \mid U(t_j, t_{j-1}) \mid q_{j-1}\rangle \approx \exp\left(-\epsilon L(p_j, q_j)\right).$$

We define the operator p as

$$\frac{\partial}{\partial q} \rightarrow i\hat{p},$$

its eigenfunctions $\langle p_j \mid$ fulfilling

$$\langle p_j \mid q_j\rangle = \frac{1}{\sqrt{2\pi}} \exp\left(ip_j q_j\right),$$

and

$$\int dp_j \mid p_j\rangle\langle p_j \mid = 1.$$

We also have that

$$f(q) \mid q'\rangle = f(q') \mid q'\rangle,$$

as well as

$$\exp\left(\epsilon \frac{\partial}{\partial q} f(q)\right) \mid p\rangle \rightarrow \exp\left(\epsilon i p f(q)\right) \mid p\rangle,$$

$$\exp\left(\epsilon D \frac{\partial^2}{\partial q^2}\right) \mid p\rangle \rightarrow \exp\left(-\epsilon D p^2\right) \mid p\rangle.$$

According to these results, each contribution to the propagator could be written as

$$\begin{aligned}
\langle q_j \mid U(t_j, t_{j-1}) \mid q_{j-1}\rangle &= \int dp_{j-1} \langle q_j \mid U(t_j, t_{j-1}) \mid p_{j-1}\rangle\langle p_{j-1} \mid q_{j-1}\rangle \\
&= \int dp_{j-1} \langle q_j \mid \exp\left(\epsilon \frac{\partial}{\partial q_j} f(q_j) - \epsilon D \frac{\partial^2}{\partial q_j^2}\right) \\
&\qquad \times \mid p_{j-1}\rangle\langle p_{j-1} \mid q_{j-1}\rangle \\
&= \int dp_{j-1} \langle q_j \mid p_{j-1}\rangle \exp\left(\epsilon[ip_{j-1} f(q_j) - D p_{j-1}^2]\right) \\
&\qquad \times \frac{1}{\sqrt{2\pi}} \exp\left(ip_{j-1} q_{j-1}\right) \\
&= \frac{1}{2\pi} \int dp_{j-1} \exp\left(i\epsilon p_{j-1}[f(q_j) - \frac{\triangle q_j}{\epsilon}] - \epsilon D p_{j-1}^2\right),
\end{aligned} \tag{2.27}$$

with $\triangle q_j = q_j - q_{j-1}$.

Introducing this expression into (2.26) we obtain

$$\langle q \mid U(t,t_0) \mid q_0 \rangle = \int \dots \int \prod_{j=1}^{M} dq_j \prod_{k=1}^{M+1} \frac{dp_k}{2\pi}$$

$$\times \prod_{j=1}^{M+1} \exp\left(i\epsilon p_{j-1}[f(q_j) - \frac{\triangle q_j}{\epsilon}] - \epsilon D p_{j-1}^2 \right)$$

$$= \int \mathcal{D}[q(t)]\mathcal{D}[p(t)] \exp\left(-\frac{1}{D} S[q,p] \right), \qquad (2.28)$$

where the last line indicates the limit of $M \to \infty$. Here, the integration measures are such that

$$\mathcal{D}[q(t)] \approx \prod_{j=1}^{M} dq_j$$

and

$$\mathcal{D}[p(t)] \approx \prod_{k=1}^{M+1} \frac{dp_k}{2\pi}.$$

We have defined

$$S[q,p] = \int ds \Big\{ ip(s)\dot{q}(s) - \mathcal{H}(q(s),p(s)) \Big\}$$

with

$$\mathcal{H}(q(s),p(s)) = -ip(s)f(q(s)) - (ip(s))^2.$$

Equation (2.28) corresponds to the phase-space representation for the propagator, a form that we will exploit in some cases. Clearly, the Gaussian integrals over p_j can be readily done, yielding the different contributions in (2.16), and leading us to the usual configuration space path integral.

2.5 Generating Functional and Correlations

We now define the so called *Generating Functional* and its relation with the correlation functions. Starting from the form defined in (2.19), we define the following form of the action

$$\mathcal{S}_J[q(t)] = \int_{t_0}^{t} ds \Big[\mathcal{L}[q(s),\dot{q}(s)] + DJ(s)q(s) \Big]. \qquad (2.29)$$

Introducing this form into the path integral we have the following functional

$$\mathcal{Z}[J(t)] = \int \mathcal{D}[q] \exp\left\{-\frac{1}{D}\mathcal{S}_J[q(t)]\right\}, \tag{2.30}$$

this functional is called the *Generating Functional.* The reason for this name can be understood as follows. Let us take the functional derivatives of $\mathcal{Z}[J(t)]$ respect to $J(t)$. What we have is

$$\frac{\delta}{\delta J(t_1)}\mathcal{Z}[J(t)] = -\int \mathcal{D}[q]\, q(t_1) \exp\left\{-\frac{1}{D}\mathcal{S}_J[q(t)]\right\}$$

$$\frac{\delta^2}{\delta J(t_1)\delta J(t_2)}\mathcal{Z}[J(t)] = \int \mathcal{D}[q]\, q(t_1)q(t_2) \exp\left\{-\frac{1}{D}\mathcal{S}_J[q(t)]\right\}$$

$$... = ...$$

$$\frac{\delta^n}{\delta J(t_1)...\delta J(t_n)}\mathcal{Z}[J(t)] = (-1)^n \int \mathcal{D}[q]\, q(t_1)...q(t_n) \exp\left\{-\frac{1}{D}\mathcal{S}_J[q(t)]\right\}.$$

When these expressions are evaluated at $J = 0$ we obtain

$$\left.\frac{\delta}{\delta J(t_1)}\mathcal{Z}[J(t)]\right|_{J=0} = -\int \mathcal{D}[q]\, q(t_1) \exp\left\{-\frac{1}{D}\mathcal{S}[q(t)]\right\}$$

$$... \quad ...$$

$$= -\langle q(t_1)\rangle$$

$$\left.\frac{\delta^2}{\delta J(t_1)\delta J(t_2)}\mathcal{Z}[J(t)]\right|_{J=0} = \int \mathcal{D}[q]\, q(t_1)q(t_2) \exp\left\{-\frac{1}{D}\mathcal{S}[q(t)]\right\}$$

$$... \quad ...$$

$$= -\langle q(t_1)q(t_2)\rangle,$$

$$... = ...$$

$$\left.\frac{\delta^n}{\delta J(t_1)...\delta J(t_n)}\mathcal{Z}[J(t)]\right|_{J=0} = (-1)^n \int \mathcal{D}[q]\, q(t_1)...q(t_n) \exp\left\{-\frac{1}{D}\mathcal{S}[q(t)]\right\}$$

$$... \quad ...$$

$$= -\langle q(t_1)...q(t_n)\rangle, \tag{2.31}$$

where $\langle ... \rangle$ indicates the average of the inserted quantity. It is clear that such averages correspond to the one, two, n-time correlation functions. In general we could define

$$\langle q(t_1)...q(t_n)\rangle = (-1)^n \frac{\delta^n}{\delta J(t_1)...\delta J(t_n)} \mathcal{Z}[J(t)]\Bigg|_{J=0}, \tag{2.32}$$

that makes clear the reason for the name of $\mathcal{Z}[J(t)]$.

This functional also allows us to obtain the average of a function $\mathcal{G}[q(t)]$, if it accepts the following expansion

$$\mathcal{G}[q(t)] = \sum_{n=0}^{\infty} G_n q(t)^n. \tag{2.33}$$

Hence, we can define its average as

$$\begin{aligned}\langle \mathcal{G}[q(t)]\rangle &= \int \mathcal{D}[q]\mathcal{G}[q(t)] \exp\left\{-\frac{1}{D}\mathcal{S}[q(t)]\right\} \\ &= \sum_{n=0}^{\infty} G_n \int \mathcal{D}[q] q(t)^n \exp\left\{-\frac{1}{D}\mathcal{S}[q(t)]\right\},\end{aligned} \tag{2.34}$$

and could be finally written as

$$\langle \mathcal{G}[q(t)]\rangle = \mathcal{G}\left[\frac{\delta}{\delta J(t)}\right] \mathcal{Z}[J(t)]\Bigg|_{J=0}. \tag{2.35}$$

Chapter 3

Generalized Path Expansion Scheme I

3.1 Expansion Around the Reference Path

The scheme we describe here is due to Davison (1954), and has been generalized by Levit and Smilansky (1985). Originally, this scheme was applied to the quantum case and we will adapt it here to the present stochastic framework.

Let $q_r(t)$ be a reference trajectory, which is an arbitrary function of t, satisfying the conditions

$$\begin{aligned} q_r(t_0) = q(t_0) &= q_0 \\ q_r(t_f) = q(t_f) &= q_f. \end{aligned} \tag{3.1}$$

Let $\{u^\nu(t)\}$ be an infinite orthogonal basis which spans the space of functions defined in the interval $[t_0, t_f]$ and vanishing at both its ends (i.e.: $u^\nu(t_o) = u^\nu(t_f) = 0$). This set can be conveniently chosen as the normalized eigenfunctions of an arbitrary Sturm–Liouville differential operator defined in this space. However we will see that just the problem under study provides the form of the most adequate operator. A completely arbitrary path can be split and written as

$$q(t) = q_r(t) + \sum_{\nu=1}^{\infty} a_\nu \, u^\nu(t). \tag{3.2}$$

Even though the reference trajectory $q_r(t)$ could be arbitrarily chosen, for convenience we select the one that makes extreme the action (2.19). The first variation of the action gives

$$\delta\mathcal{S}[q(t)] = \int_{t_0}^{t} ds \left\{ \frac{\partial \mathcal{L}}{\partial q}\,\delta q + \frac{\partial \mathcal{L}}{\partial \dot{q}}\,\delta \dot{q} \right\}$$
$$= \int_{t_0}^{t} ds \left\{ \frac{\partial \mathcal{L}}{\partial q} - \frac{d}{dt}\frac{\partial \mathcal{L}}{\partial \dot{q}} \right\} \delta q, \tag{3.3}$$

where the variations fulfill $\delta q(t_0) = \delta q(q_f) = 0$. Then, the reference (sometimes called classical, or in the present context "most probable" (Langouche *et al.* (1982))) trajectory, chosen as the one that makes extreme the action, implies $\delta\mathcal{S}[q(t)] = 0$. Hence, it is the solution of the corresponding Euler–Lagrange equation

$$\frac{d}{dt}\frac{\partial \mathcal{L}}{\partial \dot{q}} - \frac{\partial \mathcal{L}}{\partial q} = 0. \tag{3.4}$$

We can expand the action about this classical path obtaining to the lowest order

$$\mathcal{S}[q(t)] = \mathcal{S}[q_r(t)] + \frac{1}{2}\delta^2\mathcal{S}[q_r(t)]. \tag{3.5}$$

The second variation of $\mathcal{S}[q(t)]$, indicated by $\delta^2\mathcal{S}[q_r(t)]$ leads to

$$\delta^2\mathcal{S}[q_r(t)] = \int_{t_0}^{t} ds \left\{ \delta q \frac{\partial^2 \mathcal{L}}{\partial q^2}\delta q + \delta q \frac{\partial^2 \mathcal{L}}{\partial q \partial \dot{q}}\,\delta \dot{q} \right.$$
$$\left. + \delta \dot{q} \frac{\partial^2 \mathcal{L}}{\partial \dot{q} \partial q}\,\delta q + \delta \dot{q} \frac{\partial^2 \mathcal{L}}{\partial \dot{q}^2}\,\delta \dot{q} \right\}_{q_r(t)}, \tag{3.6}$$

where all the second order derivatives must be evaluated along the classical path. We call

$$\mathcal{R}[q_r, \dot{q}_r] = \left(\frac{\partial^2 \mathcal{L}}{\partial q^2} \right)_{q_r} \tag{3.7}$$

$$\mathcal{Q}[q_r, \dot{q}_r] = \left(\frac{\partial^2 \mathcal{L}}{\partial q \partial \dot{q}} \right)_{q_r} \tag{3.8}$$

$$\tilde{\mathcal{Q}}[q_r, \dot{q}_r] = \left(\frac{\partial^2 \mathcal{L}}{\partial \dot{q} \partial q} \right)_{q_r} \tag{3.9}$$

$$\mathcal{P}[q_r, \dot{q}_r] = \left(\frac{\partial^2 \mathcal{L}}{\partial \dot{q}^2} \right)_{q_r}. \tag{3.10}$$

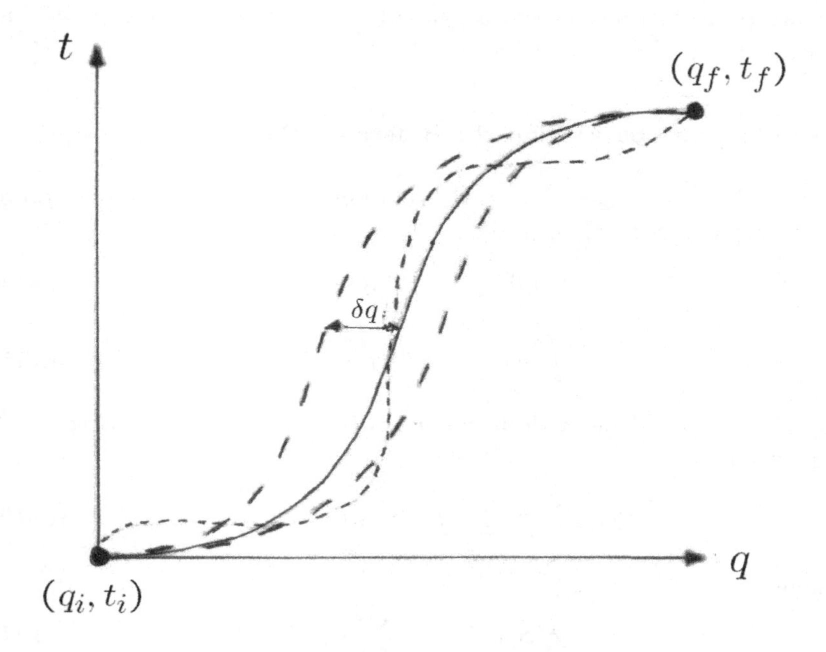

Fig. 3.1 Departure from the reference trajectory.

Now, we rewrite (3.6) as

$$\delta^2 \mathcal{S}[q_r(t)] = \int_{t_0}^{t} ds \, \left\{ \delta q \mathcal{R} \delta q + \delta q \mathcal{Q} \delta \dot{q} + \delta \dot{q} \tilde{\mathcal{Q}} \delta q + \delta \dot{q} \mathcal{P} \delta \dot{q} \right\}, \tag{3.11}$$

and after integrating by parts (once only), we obtain

$$\begin{aligned} \delta^2 \mathcal{S}[q_r(t)] &= \int_{t_0}^{t} ds \; \delta q \left\{ \mathcal{R} + \mathcal{Q} \frac{d}{dt} - \frac{d}{dt} \tilde{\mathcal{Q}} - \frac{d}{dt} \mathcal{P} \frac{d}{dt} \right\} \delta q \\ &= \int_{t_0}^{t} ds \; \delta q \Lambda \delta q. \end{aligned} \tag{3.12}$$

The explicit form of the operator Λ, also called Jacobi equation (Goldstein (1980)), is

$$\Lambda = - \left\{ \frac{d}{dt} \left(\frac{\partial^2 \mathcal{L}}{\partial \dot{q}^2} \right)_{q_r} \frac{d}{dt} + \frac{d}{dt} \left(\frac{\partial^2 \mathcal{L}}{\partial \dot{q} \partial q} \right)_{q_r} - \left(\frac{\partial^2 \mathcal{L}}{\partial q \partial \dot{q}} \right)_{q_r} \frac{d}{dt} - \left(\frac{\partial^2 \mathcal{L}}{\partial q^2} \right)_{q_r} \right\}, \tag{3.13}$$

and we reiterate that the coefficients obtained by derivatives of the Lagrangian should be evaluated at $q_r(t)$.

3.2 Fluctuations Around the Reference Path

We now choose the basis $u^\nu(t)$ as the solutions of the Sturm–Liouville problem associated with the operator Λ

$$\Lambda u^\nu(t) + \lambda^\nu u^\nu(t) = 0, \tag{3.14}$$

$$\int_{t_0}^{t} ds\ u^\nu(t)\ u^\eta(t) \ = \ \delta_{\nu\eta}, \tag{3.15}$$

with eigenvalues λ^ν and with the boundary conditions $u^\nu(t_0) \ = \ u^\nu(t) \ = \ 0$. Then we have

$$\delta^2 \mathcal{S}[q_r(t)] \ = \ \sum_{\nu,\eta} \int_{t_0}^{t} ds\ a_\nu\, a_\eta\, u^\nu(t) \Lambda u^\eta(t), \tag{3.16}$$

leading to

$$\delta^2 \mathcal{S}[q_r(t)] \ = \ \sum_{\nu} a_\nu^2\, \lambda^\nu. \tag{3.17}$$

This result makes obvious the reason for choosing the basis $\{u^\nu(t)\}$ as was done in (3.14).

Returning to (2.20) and taking into account the expansion (3.5) and the last results, we obtain

$$\begin{aligned} P(q,t \mid q_0,t_0) &\cong \int \mathcal{D}[q(t)]\ e^{-\ \mathcal{S}[q_r(t)] - \frac{1}{2}\delta^2 \mathcal{S}[q_r(t)]} \\ &\simeq e^{-\ \mathcal{S}[q_r(t)]/D} \int_{\delta q(t_0) = \delta q(t_f) = 0} \mathcal{D}[\delta q(t)]\ e^{-\ \frac{1}{2}\delta^2 \mathcal{S}[q_r(t)]} \\ &\simeq e^{-\ \mathcal{S}[q_r(t)]/D}\ \mathcal{K}(t_f \mid t_0). \end{aligned} \tag{3.18}$$

The procedure leading to (3.18) is equivalent to the usual stationary-phase approximation in the path-integral picture of standard quantum mechanics (Schulman (1981); Langouche *et al.* (1982); Levit and Smilansky (1985); Kleinert (1990b); Khandekar and Lawande (1986); Khandekar *et al.* (2000)). The first term represents the naive (classical) approximation for the propagator, while the last integral includes the fluctuations around the classical (or also deterministic) trajectory. What remains is to evaluate the integral giving $\mathcal{K}(t_f \mid t_0)$, but transforming from $\{\delta q(t)\}$ to the set $\{a_\nu\}$. Since this is a linear transformation, and thus independent of the a_ν, what remains are several independent Gaussian integrals

$$\mathcal{K}(t_f \mid t_0) = \int_\lambda \prod_\nu da_\nu \; \mathcal{J} \, e^{\sum_\nu \lambda^\nu a_\nu^2}$$

$$= \mathcal{J} \prod_\nu \left(\frac{2\pi D}{\lambda^\nu} \right)^{1/2} . \tag{3.19}$$

The Jacobian $\mathcal{J}$ must be invariant, in such a way that

$$\mathcal{K}(t_f \mid t_0) \prod_\nu \; (\lambda^\nu)^{1/2} \;=\; \mathcal{K}(t_f \mid t_0)^{free} \prod_\nu \left(\lambda^\nu_{free}\right)^{1/2} , \tag{3.20}$$

where *free* means the case without drift (that is $f(q) = 0$). In such a case the Sturm–Liouville operator takes the simple form

$$\Lambda^{free} \;=\; -\frac{d}{dt}\mathcal{P}\frac{d}{dt} \;=\; -\frac{d^2}{dt^2}. \tag{3.21}$$

From (3.19) and (3.20) it is clear how we can proceed in order to obtain approximate results for the propagator. If we only consider the set that includes the first N functions $u^\nu(t)$, we have an Nth-order approximation to the propagator through

$$P(q,t \mid q_0,t_0)_N \;=\; e^{-\,\mathcal{S}[q_r(t)]} \left(\prod_\nu^N \frac{\lambda^\nu_{free}}{\lambda^\nu} \right)^{1/2} \mathcal{K}(t_f \mid t_0)^{free}, \tag{3.22}$$

which is equivalent to the usual polygonal approximation to the path-integral (Feynman and Hibbs (1965); Schulman (1981); Langouche *et al.* (1982); Wio (1990); Khandekar and Lawande (1986); Khandekar *et al.* (2000)). This could be a good starting point for numerical procedures. However, what we want is to find more general results.

Before continuing with this calculation we now show a different result (initially using a quite simple example) related to the exploitation of space-time transformations, that will allow us to go a little further in this approach.

Chapter 4

Space-Time Transformation I

4.1 Introduction

One aspect recently studied by several authors concerns the application of space-time transformation, within the realm of path integral schemes, in order to "map" a "difficult" (in principle unsolvable) problem into a "more simple" (solvable) one. One of the most outstanding examples is the possibility of solving the Coulomb problem, within a path integral framework, via the so called *Duru–Kleinert* transformations (Duru and Kleinert (1979, 1982); Ho and Inomata (1982); Duru (1983); Inomata (1983, 1984); Pak and Sokman (1984a,b); Kleinert (1989a,b, 1990a,b); Pelster and Kleinert (1997)). However, the use of these transformations within the realm of the path integral approach to stochastic processes is scarce or almost nonexistent (Deininghaus and Graham (1979)).

In this chapter we present a simple application of such a kind of transformation for the case of diffusion in a time-dependent harmonic potential. It is well known that a closed expression exists for the transition probability of such a system (Hänggi and Thomas (1975); van Kampen (2004)). Also, a very thorough study of the most general Gaussian path integral form can be found in Grosjean (1988). However, our aim is to show, through such a simple example, the possibility of exploiting these transformation techniques, within a path integral framework, in more complicated cases. We will follow the procedure presented in Felsager (1985) for the quantum case, translating it to the stochastic (i.e. *imaginary time*) case. In what follows we present the procedure to be used, and the way to get the *classical* (or *most probable*) trajectories necessary to write the general solution (that is reduced to quadratures). In so doing, we will follow the paper of Batista *et al.* (1996). We also discuss the application of the previous results to the problem of a *fluctuation theorem* within an out of equilibrium Onsager–Machlup approach (Deza *et al.* (2009)).

4.2 Simple Example

Our starting point is to consider the following Langevin equation

$$\dot{x}(t) = h(x,t) + \xi(t), \tag{4.1}$$

where $\xi(t)$, as usual, is an additive *white noise* (van Kampen (2004)). It is well known that in one dimension a multiplicative noise problem can be reduced to an additive one (van Kampen (2004); Gardiner (2009)). Equation (4.1) corresponds to describing the overdamped motion of a particle in a time-dependent potential. In the present case we assume that $h(x,t) = -a(t)x$. As indicated, in Chapter 3, the path integral representation of the transition probability associated with this Langevin equation is given by:

$$P(x_b, t_b \mid x_a, t_a) = \int_{x(t_a)=x_a}^{x(t_b)=x_b} \mathcal{D}\left[x(t)\right] e^{-\int_{t_a}^{t_b} L(x(\tau),\dot{x}(\tau),\tau)\,d\tau}. \tag{4.2}$$

Here the stochastic *Lagrangian* or *Onsager–Machlup* functional is given, in a midpoint ($\alpha = 1/2$) discretization (Langouche *et al.* (1982)), by

$$L(x,\dot{x},t) = \frac{1}{2D}\left(\dot{x} - h(x,t)\right)^2 + \frac{1}{2}\frac{dh(x,t)}{dx}. \tag{4.3}$$

Replacing the actual form of $h(x,t)$, the previous expression can be expanded to yield:

$$L(x,\dot{x},t) = \frac{1}{2D}\left(\dot{x}^2 + a^2x^2 + 2\,a\,x\,\dot{x}\right) + \frac{1}{2}\,a = L_o + \frac{d\Phi(t)}{dt}, \tag{4.4}$$

$$\Phi(t) = -\frac{1}{2}\int_{t_a}^{t} a(t')\,dt' + \frac{a(t)}{2D}\,x^2, \tag{4.5}$$

$$L_o = \frac{1}{2D}\left(\dot{x}^2 + \left(a^2(t) - \dot{a}(t)\right)x^2\right). \tag{4.6}$$

At this point, as the Lagrangian of our problem is at most quadratic in x and $\dot{x}$, we can obtain the exact result through the use of the usual procedure of expanding around a reference (*classical* or *most probable*) trajectory ($x(t) = x_{clas}(t) + q(t)$) and, apart from the problem of getting such a trajectory, our problem reduces to evaluating the following expression:

$$P(x_b, t_b \mid x_a, t_a) = e^{-[\Phi(t_b)-\Phi(t_a)]}\, e^{-S_o^{cl}\,(t_b,t_a)} \int_{q(t_a)=0}^{q(t_b)=0} \mathcal{D}\,[q(t)]\, e^{-\delta^2 S_o(t_b,t_a)}. \tag{4.7}$$

The effective action to be solved in order to perform the path integral indicated in (4.7) is then

$$\delta^2 S_o(t_b, t_a) = \frac{1}{2D}\int_{t_a}^{t_b} (\dot{q}^2 + (a^2 - \dot{a})q^2)\, dt. \tag{4.8}$$

After partial integration, we get

$$\delta^2 S_o(t_b, t_a) = -\frac{1}{2D}\int_{t_a}^{t_b} (q\ddot{q} - (a^2 - \dot{q})q^2\, dt. \tag{4.9}$$

Hence, the path integral in (4.7) adopts the form

$$\begin{aligned} P(x_b, t_b \mid x_a, t_a) = \; & e^{-[\Phi(t_b)-\Phi(t_a)]}\, e^{-S_o^{cl}\,(t_a,t_b)} \\ & \int_{q(t_a)=0}^{q(t_b)=0} \mathcal{D}[q(t)] \exp\left[\frac{1}{2D}\int_{t_a}^{t_b} dt\, q(\frac{d^2}{dt^2} - (a^2 - \dot{a}))q\right]. \end{aligned} \tag{4.10}$$

In order to evaluate this path integral we need to diagonalize the operator

$$\frac{d^2}{dt^2} - \left[a^2 - \dot{a}\right]. \tag{4.11}$$

However, there is an alternative way involving a change of variables that transforms the action in (4.8) into the one corresponding to the free diffusion problem. In following this approach we will make use of results in Chapter 5 of Felsager (1985). Let us call $f(t)$ the solution of the equation

$$\left\{\frac{d^2}{dt^2} - w(t)\right\}\, f(t) = 0, \tag{4.12}$$

where $w(t) = a^2 - \dot{a}$, and with the condition $f(t_a) \neq 0$. It is easy to verify that

$$f(t) = A\, e^{\{-\int_{t_a}^{t}\, a(s)\, ds\}}, \tag{4.13}$$

is a solution of (4.12) fulfilling the required condition. We will use now the function $f(t)$ to perform the change of variables $q(t) \Rightarrow y(t)$ according to

$$q(t) = \; f(t) \int_{t_a}^{t} \frac{\dot{y}(s)}{f(s)}\, ds, \tag{4.14}$$

with the condition $y(t_a) = 0$. Differentiating the previous expression we get

$$\dot{q}(t) = \dot{f}(t) \int_{t_a}^{t} \frac{\dot{y}(s)}{f(s)}\, ds + \dot{y}(t) = \frac{\dot{f}(t)}{f(t)}\, q(t) + \dot{y}(t), \tag{4.15}$$

that can be inverted yielding

$$y(t) = q(t) - \int_{t_a}^{t} \frac{\dot{f}(s)}{f(s)}\, q(s)\, ds. \tag{4.16}$$

Differentiating (4.15) once more we obtain

$$\ddot{q}(t) = \ddot{f}(t) \int_{t_a}^{t} \frac{\dot{y}(s)}{f(s)}\, ds + \frac{\dot{f}(t)\, \dot{y}(t)}{f(t)} + \ddot{y}(t). \tag{4.17}$$

Replacing the last result into the integrand of the exponent in (4.10), it can be transformed into

$$\left\{\frac{d^2}{dt^2} - w(t)\right\} q(t) = \left\{\ddot{f}(t) - w(t) f(t)\right\} \int_{t_a}^{t} \frac{\dot{y}(s)}{f(s)}\, ds + \frac{\dot{f}(t)\, \dot{y}(t)}{f(t)} + \ddot{y}(t). \tag{4.18}$$

Due to (4.12), the first term on the r.h.s. of (4.18) vanishes, reducing the effective action in (4.8) to

$$\delta^2 S_o\, [q(t)] = -\frac{1}{2D} \int_{t_a}^{t_b} dt\, \left\{F(t)\, \dot{f}(t)\, \dot{y}(t) + F(t)\, f(t)\, \ddot{y}(t)\right\}, \tag{4.19}$$

where

$$F(t) = \int_{t_a}^{t} ds\, \frac{\dot{y}(s)}{f(s)}. \tag{4.20}$$

The second term on the r.h.s. of (4.19) can be integrated by parts leading to

$$\delta^2 S_o\, [q(t)] = \frac{1}{2D} \int_{t_a}^{t_b} dt\, \left(\frac{dy}{dt}\right)^2 - \frac{1}{2D} \Big[q(t)\dot{y}(t)\Big]_{t_a}^{t_b}. \tag{4.21}$$

Due to the boundary conditions at $t = t_a$ and $t = t_b$, the second term of the last equation vanishes and we finally arrive at the action corresponding to free diffusion. The boundary conditions $q(t_a) = q(t_b) = 0$, when written in terms of the new variable $y(t)$, have the form:

$$y(t_a) = 0 \quad ; \quad \int_{t_a}^{t_b} ds\, \frac{\dot{y}(s)}{f(s)} = 0. \tag{4.22}$$

However, the second boundary condition is a nonlocal one and therefore we shall resort to a special trick in order to handle it. Such a trick involves using the identity $\delta(q(t_b)) = (2\pi)^{-1} \int_{-\infty}^{\infty} \exp[-isq(t_b)]\, ds$, in order to formally introduce the integration over the final end-point

$$\int_{q(t_a)=0}^{q(t_b)=0} \mathcal{D}[q]\, e^{-\delta^2 S_o[q(t)]} = \frac{1}{2\pi} \int_{q(t_a)=0}^{q(t_b)\ arbitrary} \mathcal{D}[q] \times \int_{-\infty}^{\infty} ds\, e^{-is\, q(t_b)}\, e^{-\delta^2 S_o[q(t)]}, \tag{4.23}$$

where the integration over s produces the δ-function that takes care of the correct boundary condition. Changing the integration variables $(q(t) \Rightarrow y(t))$ we get

$$\frac{1}{2\pi} \int_{y(t_a)=0}^{y(t_b)\ arbitrary} D[y(t)] \int_{-\infty}^{\infty} ds \exp\left[-is\left(f(t_b) \int_{t_a}^{t_b} \frac{\dot{y}(s')}{f(s')}\, ds'\right)\right] \times \exp\left[-\frac{1}{2D} \int_{t_a}^{t_b} dt \left(\frac{dy}{dt}\right)^2\right] \det\left[\frac{\delta q}{\delta y}\right]. \tag{4.24}$$

As the transformation $q(t) \Rightarrow y(t)$ is linear, the Jacobian $det[\delta q/\delta y]$ is independent of $y(t)$, and the remaining integral is Gaussian. *Completing the square* we get

$$= \frac{1}{2\pi} \det\left[\frac{\delta q}{\delta y}\right] \int_{-\infty}^{\infty} ds \exp\left[-\frac{D}{2}\, s^2\, f^2(t_b) \int_{t_b}^{t_a} \frac{dt}{f^2(t)}\right] \times \int_{\vartheta(t_a)=0}^{\vartheta(t_b)\ arbitrary} D[\vartheta(t)] \exp\left[-\frac{1}{2D} \int_{t_a}^{t_b} dt \left(\frac{d\vartheta}{dt}\right)^2\right], \tag{4.25}$$

with

$$\vartheta(t) = y(t) - iD\, s\, f(t_b) \int_{t_a}^{t} \frac{d\alpha}{f(\alpha)}. \tag{4.26}$$

The second integral in (4.25) is equal to unity as it represents the probability for finding the free diffusive system anywhere at time t_b. Hence, integrating the first one, we get the simple expression

$$\int_{q(t_a)=0}^{q(t_b)=0} D[q]\, e^{-\delta^2 S_o[q(t)]} = \left[2\pi D\, f(t_a)\, f(t_b) \int_{t_a}^{t_b} \frac{dt}{f(t)^2}\right]^{-\frac{1}{2}}. \tag{4.27}$$

Here we have used that the Jacobian is given by $\det[\delta q/\delta y] = \sqrt{f(t_b)/f(t_a)}$ (Felsager (1985)). The final form for the transition probability is

$$P(x_b, t_b \mid x_a, t_a) = e^{-[\Phi(t_b)-\Phi(t_a)]}\, e^{-S_o^{cl}\,(t_a,t_b)} \left[2\pi D\, f(t_a) f(t_b) \int_{t_a}^{t_b} \frac{dt}{f(t)^2}\right]^{-\frac{1}{2}} . \tag{4.28}$$

Clearly, the expression for this transition probability contains the results for free diffusion ($\omega(t) = 0,\ f(t) = 1$) and diffusion in a constant harmonic potential ($\omega(t) = \omega^2 = ct.,\ f(t) = \cosh(\omega(t - t_a))$). In the general case, the function $f(t)$ is given by (4.13). It is worth remarking that the result in (4.28) is in complete agreement with those obtained by previous authors by other means (see for instance Grosjean (1988)).

In order to completely solve the problem, i.e. to have the transition probability in (4.28), we shall evaluate $S_o^{cl}\,(t_b, t_a) = S_o[x_{clas}\,(t)]$, where $x_{clas}\,(t)$ is the solution of the Euler–Lagrange equation

$$\left[\frac{d^2}{dt^2} - \left[a^2 - \dot{a}\right]\right] x_{clas}(t) \;=\; 0 \tag{4.29}$$

fulfilling the "boundary conditions":

$$x_{clas}\,(t_a) = x_a$$

and

$$x_{clas}\,(t_b) = x_b .$$

The general form of such a solution is

$$x_{clas}(t) \;=\; B\,\exp\left\{-\int_{t_a}^{t} a(s)\,ds\right\} \;+\; g(t), \tag{4.30}$$

where $g(t)$ is an independent solution of the equation of motion that shall be obtained for each $a(t)$. It is easy to check that

$$g(t) \;=\; \exp\left\{-\int_{t_a}^{t} a(s)\,ds\right\} \int_{t_a}^{t} d\tau\, \exp\left\{2\int_{t_a}^{\tau} a(z)\,dz\right\}, \tag{4.31}$$

is a convenient form of the desired independent solution. The knowledge of $x_{clas}\,(t)$ allows us to evaluate S_o^{cl}. Hence, we have reached an expression where, given the form of the time dependence of $a(t)$, the complete solution of the problem is reduced to *quadratures*. A whole family of exact solutions can be found in (Batista *et al.* (1996)).

4.3 Fluctuation Theorems from Non-equilibrium Onsager–Machlup Theory

Fluctuations around deterministic behavior, has been a prime source of information about the system's dynamics (a known example is the fluctuation–dissipation theorem established in linear response theory near equilibrium, linking transport coefficients to the autocorrelation functions of fluctuations, and evolved from Einstein's relation, Nyquist's theorem, Onsager's reciprocal relations, etc. (Wio, Deza and López (2012))), and since they can induce non-equilibrium phase transitions, lead to the appearance of spatiotemporal patterns, and assist, enhance, or sustain a host of mesoscopic phenomena impossible in their absence.

The natural framework to describe their temporal behavior is the theory of stochastic processes. A milestone was *Onsager–Machlup's fluctuation theory* around equilibrium, using a functional integral approach for stochastic linear relaxation processes, and leading to a variational principle known as Onsager's principle of minimum energy dissipation. Another milestone has been set by recent *fluctuation theorems* (Jarzynski (2008); Evans and Searles (2002); Chernyak, *et al.* (2006); van den Broeck (2010); van den Broeck *et al.* (2010)), asymmetric relations for the distribution functions of work, heat, etc. Several laboratory experiments have verified their validity even far from equilibrium or from the thermodynamic limit.

The validity of fluctuation theorems far from equilibrium poses a challenge to generalize the Onsager–Machlup theory to non-equilibrium states. Taniguchi and Cohen (2007) have succeeded in generalizing it to *steady* states out of equilibrium, by considering a harmonically bound (spring constant κ) Brownian particle which is dragged with constant speed v through a fluid (friction constant α, inverse temperature β) acting as heat reservoir. They have thus derived non-equilibrium analogs of the laws of thermodynamics, and steady-state detailed balance relations from which (among others) a fluctuation theorem for work (valid in the long-time limit) can be derived. In the overdamped case and in a co-moving system $[y(t) \equiv x(t) - vt]$, the dynamics are described by the Langevin equation

$$\dot{y} = -\frac{1}{\tau}y(t) - v + \frac{1}{\alpha}\zeta(t), \tag{4.32}$$

$\tau \equiv \alpha/\kappa$ being a relaxation time and $\zeta(t)$ a Gaussian white noise, with $\langle\zeta(t)\rangle = 0$ and $\langle\zeta(t)\zeta(t')\rangle = 2D\delta(t-t')$. In order that equilibrium be attained for $v = 0$, it must be $D \equiv \alpha/\beta$ and $\langle\ldots\rangle$ an average over the initial ensemble. Note that whereas the dynamics of (4.32) is invariant under

Fig. 4.1 An artistic view of a perpetuum mobile! (A fragment of "Waterfall", lithograph by M. C. Escher, October, 1961)

the joint transformation $y(t) \to -y(t), v \to -v$, the Gaussian white-noise property of $\zeta(t)$ does not change to $\zeta(t) \to -\zeta(t)$.

The functional integral that gives the transition probability can be completely calculated, thus obtaining the result

$$P(y,t|y_0,t_0) = \frac{1}{\sqrt{4\pi D\mathcal{T}(t)}} \exp\left\{-\frac{[y(t) + v\tau - (y_0 + v\tau)\, b(t)]^2}{4D\mathcal{T}(t)}\right\},$$

where $b(t) \equiv \exp[-(t-t_0)/\tau]$ and $\mathcal{T}(t) \equiv (\tau/2)[1 - b^2(t)]$, so that $\mathcal{T}(t) = t - t_0 + \mathcal{O}[(t-t_0)^2]$.

In the generalized Onsager–Machlup theory for non-equilibrium *steady* states, the Lagrangian can be written in the form

$$L(\dot{y}(s), y(s); v) = -\frac{1}{2k_B}\left[\Phi(\dot{y}(s); v) + \Psi(y(s)) - \dot{\mathcal{S}}(\dot{y}(s), y(s); v)\right],$$

where k_B is Boltzmann's constant. The *dissipation functions* $\Phi(\dot{y}(s); v)$ and $\Psi(y(s))$, as well as the *entropy production rate* $\dot{\mathcal{S}}(\dot{y}(s), y(s); v)$ are defined by

$$\Phi(\dot{y}(s); v) \equiv \frac{\alpha}{2T}(\dot{y}(s) + v)^2 \quad ,$$

$$\Psi(y(s)) \equiv \frac{\alpha}{2T}\left(\frac{y(s)}{\tau}\right)^2 \quad ,$$

$$\dot{\mathcal{S}}(\dot{y}(s), y(s); v) \equiv -\frac{1}{T}\kappa y(s)(\dot{y}(s) + v),$$

respectively, with the temperature $T \equiv (k_B\beta)^{-1}$. Among the main results found in Taniguchi and Cohen (2007) are a non-equilibrium detailed balance relation

$$\frac{\exp\left[\int_{t_0}^{t} ds\ L\left(\dot{y}(s), y(s); v\right)\right] f_{eq}(y_0)}{\exp\left[\int_{t_0}^{t} ds\ L\left(-\dot{y}(s), y(s); -v\right)\right] f_{eq}(y(t))} = \exp\left[\beta \mathcal{W}_t(\{y(s)\}; v)\right],$$

and an *asymptotic* fluctuation theorem for work

$$\lim_{t \to +\infty} \frac{P_w(W, t)}{P_w(-W, t)} = \exp(W).$$

Taniguchi and Cohen's approach was extended to non steady states (Singh (2008)), by submitting a free Brownian particle with charge q to an external uniform oscillating field with strength E_0. The corresponding Langevin equation is

$$\dot{x} = -\eta \cos \omega t + \frac{1}{\alpha}\xi(t),$$

with $\eta \equiv qE_0/\alpha$, $\langle \xi(t) \rangle = 0$, $\langle \xi(t)\xi(t') \rangle = 2D\delta(t - t')$, and $D \equiv \alpha/\beta$. Again, the functional integral that gives the transition probability can be completely calculated

$$P(x, t|x_0, 0) = \frac{1}{\sqrt{4\pi Dt}} \exp\left[-\frac{1}{4Dt}\left(x(t) - x_0 + \frac{\eta}{\omega}\sin \omega t\right)^2\right],$$

yielding a work fluctuation theorem valid for all t.

4.4 Brownian Particle in a Time-Dependent Harmonic Potential

Here we explore the consequences of forcing in a *multiplicative* form a harmonically bound Brownian particle. Namely, we consider it as bound by $V(x, t) = \frac{1}{2}a(t)x^2$, with $a(t)$ externally controlled.

We start from the Langevin equation $\dot{x} = h(x,t) + \xi(t)$, with a drift term of the form $h(x,t) = -a(t)x$. This is exactly (4.1).

As shown before, we could write for the conditional probability

$$P(x_b, t_b|x_a, t_a) = \exp\{-[\Phi(t_b) - \Phi(t_a)]\} \exp\left[-S_0^{cl}(x_a, x_b)\right] \times \left[2\pi D f(t_a) f(t_b) \int_{t_a}^{t_b} \frac{ds}{f^2(s)}\right]^{-1/2} . \quad (4.33)$$

A simple method has been proposed in Batista *et al.* (1996) to generate analytically a whole family of classical solutions. It consists in writing the elastic parameter in the form $a(t) = \pm b(t) + \frac{1}{2}\dot{b}(t)/b(t)$, with $b(t)$ such that $B(t) = \int_{t_0}^{t} b(s)ds$ exists. Thus for $C = 1$, the aforementioned solutions $f(t)$ can be written as $x_1(t) = \sinh[B(t)]/\sqrt{b(t)}$ and $x_2(t) = \cosh[B(t)]/\sqrt{b(t)}$. This readily enables us to find the general solution with boundary conditions $x_{cl}(t_a) = x_a$ and $x_{cl}(t_b) = x_b$

$$x_{cl} = \frac{1}{\sinh[B(t_b) - B(t_a)]\sqrt{b(t)}} \left(x_a\sqrt{b(t_a)}\ x_b\sqrt{b(t_b)}\right) \times \begin{pmatrix} \sinh B(t_b) & -\cosh B(t_b) \\ -\sinh B(t_a) & \cosh B(t_a) \end{pmatrix} \begin{pmatrix} \cosh B(t) \\ \sinh B(t) \end{pmatrix}.$$

Whereas it is straightforward to rewrite $\Phi(t)$ in terms of $b(t)$ and $B(t)$, yielding

$$\exp\{-[\Phi(t_b) - \Phi(t_a)]\} =$$

$$= \exp\left[\frac{B(t_b) - B(t_a)}{2}\right] \left[\frac{b(t_b)}{b(t_a)}\right]^{1/4} \exp\left\{\frac{1}{2D}\left[a(t_b)x_b^2 - a(t_a)x_a^2\right]\right\},$$

the remaining two factors in (4.33) are not equally easy to calculate. The results are

$$S_0^{cl}(x_a, x_b) = \frac{1}{2D}\int_{t_a}^{t_b} \left[\dot{x}_{cl}^2 + \omega^2(s)x_{cl}^2\right] ds = \frac{1}{2D}\left[x_b\dot{x}(t_b) - x_a\dot{x}(t_a)\right]$$

and

$$\int_{t_a}^{t_b} \frac{ds}{x_{cl}^2(s)} = \frac{2\sinh[B(t_b) - B(t_a)]}{\sqrt{b(t_a)b(t_b)}\, x_a\, x_b},$$

giving as the final result

$$P(x_b, t_b|x_a, t_a) = \left\{ \frac{4\pi D \sinh[B(t_b) - B(t_a)]}{\sqrt{b(t_a)b(t_b)}} \right\}^{-1/2} \left[\frac{b(t_b)}{b(t_a)} \right]^{1/4}$$

$$\times \exp\left[\frac{B(t_b) - B(t_a)}{2} \right] \exp\left\{ \frac{1}{2D} \left[a(t_b)x_b^2 - a(t_a)x_a^2 \right] \right\}$$

$$\times \exp\left\{ -\frac{1}{2D} \left[x_b \dot{x}(t_b) - x_a \dot{x}(t_a) \right] \right\}. \tag{4.34}$$

4.5 Work Distribution Function

Following Taniguchi and Cohen (2007) and Singh (2008), we denote by

$$\beta W_t(\{x(s)\})$$

the dimensionless work made at time t by the Brownian particle on the stochastic trajectory $x(s)$, namely

$$W_t(\{x(s)\}) = \int_{t_a}^{t} \dot{W}(x(s), s) ds.$$

Since this is made against the drift force $-a(s)x(s)$, the power developed at time s on $x(s)$ is $\dot{W}(x(s), s) = -a(s)x\dot{x}$.

The dimensionless work distribution function W is a sum over trajectories [included a sum over x_b and an average over x_a with distribution $f_{inic}(x_a, t_a)$] with the constraint $W = \beta W_t(\{x(s)\})$ (the work over *that* trajectory) (Singh (2008))

$$P_W(W, t) = \int dx_b \int dx_a\, f_{inic}(x_a, t_a) \int \mathcal{D}[x(t)] e^{\int_{t_a}^{t} L(s)ds} \delta[W - \beta W_t(\{x(s)\})]$$

Using the delta-function representation

$$\delta[W - \beta W_t(\{x(s)\})] = \frac{1}{2\pi} \int_{-\infty}^{\infty} d\lambda\, e^{i\lambda[W - \beta W_t(\{x(s)\})]}$$

we arrive at an effective Lagrangian $L_{eff}(x, \dot{x}, t; \lambda) = L_0 - \lambda\beta\dot{W}(\{x(s)\})$, that in our case is $L_{eff} = L_0 + \frac{d\Phi}{dt} + \lambda\beta a(t)x\dot{x}$ and can be cast as

$$L_{eff}(x, \dot{x}, t; \lambda) = \frac{d\Psi}{dt} + \frac{1}{2D} \left[\dot{x}^2 + \Omega^2(t; \lambda)x^2 \right],$$

where $\Psi(t) = \Phi(t) + \frac{\lambda\beta}{2}a(t)x^2$ and

$$\Omega^2(t; \lambda) = a(t)^2 - \left(1 + \frac{\lambda\beta}{2} \right) \dot{a}(t). \tag{4.35}$$

Then we may write

$$P_W(W,t) = \frac{1}{2\pi}\int_{-\infty}^{\infty} d\lambda\, e^{i\lambda W} \int dx_b \int dx_a f_{inic}(x_a,t_a)\,\mathcal{F}(x_b,x_a,i\lambda), \tag{4.36}$$

with

$$\mathcal{F}(x_b,x_a,i\lambda) = \int_{x_a}^{x_b} \mathcal{D}[x(t)]\, \exp\left\{\int_0^t ds[\dot{x}^2 + \Omega^2(s;\lambda)x^2]\right\}. \tag{4.37}$$

Now the *classical trajectory* $g(t,\lambda)$ is solution of

$$\left[\frac{d^2}{dt^2} - a^2 + \left(1+\frac{\lambda\beta}{2}\right)\dot{a}\right] g(t,\lambda) = 0.$$

This equation is not trivially solvable by quadratures. Moreover, the fact that one must find a solution *for each value* of the integration variable λ makes numerical resolution (by e.g. simulated annealing, genetic algorithms, etc) a titanic task. One way out this problem is to take advantage of the fact that $\mathcal{F}(x_b,x_a,i\lambda)$ is by its definition an analytic function of λ. Using (4.33) $\mathcal{F}(x_b,x_a,i\lambda)$ takes the form

$$\mathcal{F}(x_b,x_a,i\lambda) = \exp\{-[\Phi(t_b)-\Phi(t_a)]\}\exp\left\{-\frac{\lambda\beta}{2}\left[a(t_b)x_b^2 - a(t_a)x_a^2\right]\right\}$$

$$\times \exp\left[-S_0^{cl}(x_a,x_b,\lambda)\right]\left[2\pi D\, g(t_a,\lambda)g(t_b,\lambda)\int_{t_a}^{t_b}\frac{ds}{g(s,\lambda)^2}\right]^{-1/2}, \tag{4.38}$$

which shows that the strong λ–dependence is in $\exp\left[-S_0^{cl}(x_a,x_b,\lambda)\right]$, whereas the prefactor (the last factor in square bracket) has a much less relevant dependence on λ. Hence, a reasonable way to advance in an analytical scheme, without committing a gross error, is to approximate the prefactor by its correspondent in (4.33), namely

$$\left[2\pi D\, g(t_a,\lambda)g(t_b,\lambda)\int_{t_a}^{t_b}\frac{ds}{g(s,\lambda)^2}\right]^{-1/2} \cong \left\{\frac{4\pi D\sinh[B(t_b)-B(t_a)]}{\sqrt{b(t_a)b(t_b)}}\right\}^{-1/2}.$$

As $S_0^{cl}(x_a,x_b,\lambda)$ is also an analytic function of λ, the following step is to expand $S_0^{cl}(x_a,x_b,\lambda)$ up to second order in λ. Now, the aforementioned expansion requires finding $\partial S_0^{cl}/\partial\lambda$ and $\partial^2 S_0^{cl}/\partial\lambda^2$, which in turn requires expanding $g(t,\lambda)$ up to second order around $\lambda=0$. In fact, from

$$S_0^{cl}(x_a,x_b,\lambda) = \int_0^t ds\left[\mathcal{L}_0 + \frac{\lambda\beta}{2}\dot{a}(t)\right]\left[g_0(t)+\lambda g_1(t)+\lambda^2 g_2(t)\right]$$

$$= S^{(0)} + \lambda S^{(1)} + \lambda^2 S^{(2)},$$

with $\mathcal{L}_0 = (d^2/dt^2) - w^2(t)$ (with $w^2(t) = a(t)^2 - \dot{a}(t)$), we obtain

$$\begin{aligned} S_0^{cl}\Big|_{\lambda=0} &= S^{(0)} = \int_{t_a}^{t_b} ds\, \mathcal{L}_0 g_0(s), \\ \frac{\partial S_0^{cl}}{\partial \lambda}\Big|_{\lambda=0} &= S^{(1)} = \int_{t_a}^{t_b} ds \left[\mathcal{L}_0 g_1(s) + \frac{\beta}{2}\dot{a}(t) g_0(s)\right], \\ \frac{1}{2}\frac{\partial^2 S_0^{cl}}{\partial^2 \lambda}\Big|_{\lambda=0} &= S^{(2)} = \int_{t_a}^{t_b} ds \left[\mathcal{L}_0 g_2(s) + \frac{\beta}{2}\dot{a}(t) g_1(s)\right] \end{aligned}$$

(note that only $S^{(0)}$ and the second term in $S^{(1)}$ can depend on x_a and x_b, since $g_1(s)$ and $g_2(s)$ have 0 as endpoints). Once $g_1(t)$ and $g_2(t)$ are found through

$$g_1(t) = -\frac{\beta}{2}\mathcal{L}_0^{-1}\dot{a}(t) g_0(t)$$

and

$$g_2(t) = \left(\frac{\beta}{2}\right)^2 \mathcal{L}_0^{-1}\dot{a}(t)\mathcal{L}_0^{-1}\dot{a}(t) g_0(t),$$

the above scheme would allow us to obtain an approximation for $P_W(W,t)$ and $P_W(-W,t)$, and evaluate the ratio

$$\frac{P_W(W,t)}{P_W(-W,t)} = \frac{\int dx_b \exp\left[-S^{(0)}(x_a,x_b)\right] \int_{-\infty}^{\infty} d\lambda \exp\left[i\lambda W - R\lambda - S^{(2)}\lambda^2\right]}{\int dx_b \exp\left[-S^{(0)}(x_a,x_b)\right] \int_{-\infty}^{\infty} d\lambda \exp\left[-i\lambda W - R\lambda - S^{(2)}\lambda^2\right]},$$

where the first and last terms in (4.38) cancel out, and we have chosen $f_{inic}(x_a,t_a) = \delta(x_a)$. The coefficient of λ is

$$R = S^{(1)}(x_a,x_b) + \frac{\beta}{2}\left[a(t_b)x_b^2 - a(t_a)x_a^2\right].$$

The integrals over λ are bound to give Gaussians of the form

$$\exp\left[(\pm W - W_{eff}(x_a,x_b,t))^2\right]$$

and the integrals over x_b should not spoil that behavior. The final expression is not included here. Meanwhile, we show that even an oversimplified calculation of $P_W(W,t)$ yields a fluctuation or work theorem, at least for $t \to \infty$. We write (4.35) as $\Omega^2(t;\lambda) = w^2(t) - \lambda\gamma(t)$ with $\gamma(t) = \left(\frac{\beta}{2}\right)\dot{a}(t)$, and expand in powers of λ the exponent of $\mathcal{F}(x_b, x_a, i\lambda)$

$$\exp\left\{\int_0^t ds[\dot{x}^2 + \Omega^2(s;\lambda)x^2]\right\} \approx$$

$$\approx \exp\left\{\int_0^t ds[\dot{x}^2 + w^2(s)x^2]\right\}\left[1 - \lambda\int_0^t ds_1\gamma(s_1)x^2(s_1)\right.$$

$$\left. + \frac{\lambda^2}{2}\int_0^t ds_1\int_0^t ds_2\gamma(s_1)x^2(s_1)\gamma(s_2)x^2(s_2) + \ldots\right] \quad (4.39)$$

By re-parameterizing the integration variable in the second and third terms of the expansion so to absorb the functions $\gamma(s)$, we are left respectively with $\sigma_2(\tau = 0) = \langle x(t)x(t)\rangle$ and $\sigma_4(\tau = 0) = \langle x^2(t)x^2(t)\rangle$. After re-summing as in (Wio *et al.* (1995)), the expression between brackets is now approximately $\exp\left[-\lambda\sigma_2(\tau = 0) + \frac{\lambda^2}{2}\sigma_4(\tau = 0)\right]$. To proceed further, we resort to $\sigma_4(\tau = 0) \approx \sigma_2^2(\tau = 0)$. Then $\sigma_2(\tau = 0)$ is found by setting $t_b = t_a$, $\theta = -1/2$ in equation (8) of Strier *et al.* (2000), namely

$$\langle x(t_a)x(t_b)\rangle = \frac{2D}{\sqrt{g(t_a)g(t_b)}}\frac{\Gamma^2(\theta + 3/2))}{\Gamma^2(\theta + 1)}\mathcal{F}(-\frac{1}{2}, -\frac{1}{2}; \theta + 1; I(t_a)/I(t_b)),$$

$\mathcal{F}$ being the hypergeometric function, $I(t) = \int_{-\infty}^t d\tau \exp[2\int_{-\infty}^{\tau} a(\xi)d\xi]$ and $g(t) = \lim_{t_0\to-\infty}\left\{\exp[2\int_{t_0}^t a(\xi)d\xi]\left(\int_{t_0}^t d\tau\exp[2\int_{t_0}^{\tau} a(\xi')d\xi']\right)^{-1}\right\}$. By performing the inverse transform we get

$$P_W(W, t) \approx \exp\left\{-\frac{W^2}{4\sigma_2(t)} + \frac{W}{2} + C(t)\right\},$$

where $C(t)$ includes all the contributions not depending upon either λ or W, and we have disregarded pre-factors and/or constants. The important fact is that even this crude approximation yields a meaningful asymptotic fluctuation theorem for work

$$\lim_{t\to\infty}\frac{P_W(W, t)}{P_W(-W, t)} = e^W. \quad (4.40)$$

Hence, we have worked out another generalization to non-equilibrium states of *Onsager–Machlup's theory*, in the spirit of those of Taniguchi and Cohen (2007); Singh (2008), deriving a fluctuation or *work* theorem, through the calculation of the work probability distribution as a functional integral over stochastic trajectories.

Chapter 5

Generalized Path Expansion Scheme II

5.1 Path Expansion: Further Aspects

The main result of Chapter 4 can be summarized as follows
Consider a general problem whose Lagrangian has the form (2.18). *The path expansion scheme, with the classical path as reference, gives the approximation* (3.18) *for the propagator. For the general Sturm–Liouville problem* (3.14) *(whose particular form was shown in* (4.11)) *we look for the solutions of the type* (4.12)

$$\Lambda f(t) = 0, \tag{5.1}$$

where $f(t)$ is any solution fulfilling $f(t_0) \neq 0$. Hence we can write for the propagator the solution of the form (4.28), *that is*

$$P(x_f, t_f \mid x_0, t_0) \propto \left[f(t_0) f(t_f) \int_{t_0}^{t_f} \frac{dt}{f(t)^2} \right]^{-\frac{1}{2}}. \tag{5.2}$$

Taking into account (3.22) and (5.2) we can write

$$\frac{\det \Lambda}{\det \Lambda^{free}} = \frac{f(t_0) f(t_f) \int_{t_0}^{t_f} dt f(t)^{-2}}{f^{free}(t_0) f^{free}(t_f) \int_{t_0}^{t_f} dt f^{free}(t)^{-2}} \tag{5.3}$$

Here, we assume that $f(t)$ and $f^{free}(t)$ do not vanish at $t = t_0$. We can simplify this last relation going to the singular limit. That is when both $f(t)$ and $f^{free}(t)$ do vanish for $t = t_0$!.

In order to prove it, let us consider the following initial value problem

$$\Lambda \beta(t) = 0, \tag{5.4}$$

with the conditions $\beta(t_0) = 0$ and $\dot{\beta}(t_0) = 1$, and the associated free problem

$$\Lambda^{free} \beta_f(t) = 0, \tag{5.5}$$

with $\beta_f(t_0) = 0$ and $\dot{\beta}_f(t_0) = 1$.

We denote by $\beta^{(1)}$ the solution of (5.4) satisfying

$$\beta^{(1)}(t_0) = 1 \qquad \frac{d}{dt}\beta^{(1)}(t_0) = 0, \tag{5.6}$$

and similarly for the associated $\beta_f^{(1)}(t)$. We can introduce in the identity (5.3) the following notation

$$\begin{aligned} f(t) &= \beta(t) + \epsilon\beta^{(1)}(t) \\ f^{free}(t) &= \beta_f(t) + \epsilon\beta_f^{(1)}(t). \end{aligned} \tag{5.7}$$

It follows that

$$\lim_{\epsilon\to 0} \frac{f(t_0)}{f^{free}(t_0)} = 1, \tag{5.8}$$

and

$$\lim_{\epsilon\to 0} \frac{f(t_f)}{f^{free}(t_f)} = \frac{\beta(t_f)}{\beta_f(t_f)}. \tag{5.9}$$

Finally, the limit $\epsilon \to 0$ of the integral

$$\int_{t_0}^{t_f} \frac{dt}{f(t)^2}, \tag{5.10}$$

diverges, because $\beta(t)$ vanishes as $t \to t_0$. But since almost all the contribution to the integral comes from the tiny neighborhood of $t = t_0$ (when $\epsilon \to 0$), this integral diverges like

$$\int_{t_0}^{t_f} \frac{dt}{(t-t_0)^2}. \tag{5.11}$$

Consequently

$$\lim_{\epsilon\to 0} \frac{\int_{t_0}^{t_f} \frac{dt}{f(t)^2}}{\int_{t_0}^{t_f} \frac{dt}{f^{free}(t)^2}} \approx 1. \tag{5.12}$$

Thus the identity (5.3) collapses into the simple determinant relation

$$\frac{\det \Lambda}{\det \Lambda^{f}ree} = \frac{\beta^{(1)}(t_f)}{\beta_f^{(1)}(t_f)}, \tag{5.13}$$

that, in the free diffusion case, reduces to

$$\frac{\det \Lambda}{\det \Lambda^{f}ree} = \frac{\beta^{(1)}(t_f)}{t_f - t_0}. \tag{5.14}$$

In the general case we find the relation

$$\left|\frac{\prod_\nu \lambda^\nu_{free}}{\prod_\eta \lambda^\eta}\right| = \left|\prod_\nu \frac{\lambda^\nu_{free}}{\lambda^\nu}\right| = \left|\frac{\det \beta^{(1)}_f(t_f)}{\det \beta^{(1)}(t_f)}\right|. \tag{5.15}$$

To prove it in general is involved, making use of some aspects of Morse's theory (we refer the interested reader to the appendix of Levit and Smilansky (1985)).

What is the physical meaning of the functions $\beta^{(1)}(t)$ and $\beta^{(1)}_f(t)$? Let $q_{class}(q_0, \dot{q}_0, t)$ be a family of classical paths for a fixed q_0, and with $\dot{q}_0$ as a parameter. Then the *response functions*

$$J(\dot{q}_0, t) = \frac{\partial q_{class}(q_0, \dot{q}_0, t)}{\partial \dot{q}_0}$$

correspond to the study of the variations of the classical path due to infinitesimal changes of $\dot{q}_0$

$$q_{class}(q_0, \dot{q}_0 + \varepsilon, t) - q_{class}(q_0, \dot{q}_0, t) = \varepsilon J(\dot{q}_0, t) + O(\varepsilon^2).$$

But $q_{class}(q_0, \dot{q}_0, t)$ satisfies the equation (3.4)

$$\frac{d}{dt}\frac{\partial \mathcal{L}}{\partial \dot{q}} - \frac{\partial \mathcal{L}}{\partial q} = 0. \tag{5.16}$$

If we differentiate this equation with respect to $\dot{q}_0(t)$, we get

$$\frac{d}{dt}\left(\frac{\partial^2 \mathcal{L}}{\partial \dot{q}^2} J(\dot{q}_0, t)\right) + \left(\frac{\partial^2 \mathcal{L}}{\partial q \partial \dot{q}} - \frac{\partial^2 \mathcal{L}}{\partial q^2}\right) J(\dot{q}_0, t) = 0, \tag{5.17}$$

that is again the Jacobi equation. But, as $q_{class}(q_0, \dot{q}_0, t = t_o) = q_o$ for all values of $\dot{q}_0$, we have $J(\dot{q}_0, t_o) = 0$.

When $J(\dot{q}_0, t_f) = 0$, we have a set of conjugate or focal points. What is important is that $J(\dot{q}_0, t)$ could be written in terms of the action. We have that (Goldstein (1980))

$$\dot{q}_0 = -\left.\frac{\partial S}{\partial q_o}\right|_{q_f, t_f}, \tag{5.18}$$

that implies

$$\frac{1}{J} = \frac{\partial \dot{q}_0}{\partial q_f} = -\left.\frac{\partial^2 S}{\partial q_o \partial q_f}\right|_{q_f, t_f}. \tag{5.19}$$

This somehow "proves" that

$$\begin{aligned} \beta^{(1)}(t) &\simeq \partial q_{class}(q_0, \dot{q}_0, t)/\partial \dot{q}_0 \\ \beta^{(1)}_f(t) &\simeq \partial q^{free}_{class}(q_0, \dot{q}_0, t)/\partial \dot{q}_0. \end{aligned} \tag{5.20}$$

The notation in (5.15) applies to the higher dimensional case, where it is necessary to evaluate the determinant of a matrix whose elements are the derivative of each component of the final coordinate with respect to every component of the initial velocity. In a quantum mechanical framework it is known as *van Vleck determinant* (Landau and Lifshitz (1958); Schulman (1981)).

Using the previous results we can rewrite (3.20) as

$$\mathcal{K}(t_f \mid t_0) = \mathcal{K}(t_f \mid t_0)^{free} \left| \frac{\partial q_{class}^{free}(q_0, \dot{q}_0, t_f)/\partial \dot{q}_0}{\partial q_{class}(q_0, \dot{q}_0, t_f)/\partial \dot{q}_0} \right|^{1/2} . \quad (5.21)$$

For the free case

$$q_{class}^{free}(t) = q_0 + \dot{q}_0(t - t_0), \quad (5.22)$$

then

$$\partial q_{class}^{free}(q_0, \dot{q}_0, t_f)/\partial \dot{q}_0 = (t - t_0), \quad (5.23)$$

and the classical prefactor in (3.18) is

$$\begin{aligned} e^{-\,\mathcal{S}[q_r(t)]/D} &= e^{-\frac{1}{2D}\int_{t_0}^{t_f} ds \dot{q}_{free}(s)^2}, \\ &= e^{-\frac{1}{2D}\dot{q}_{class,0}^2}, \\ &= e^{-\frac{(q_f - q_0)^2}{2D(t_f - t_0)}}. \end{aligned} \quad (5.24)$$

From (3.19) it is clear that

$$\mathcal{K}^{free}(t_f \mid t_0) \propto J(t_f - t_0)^{1/2}, \quad (5.25)$$

and from the normalization of $P(q, t \mid q_0, t_0)$ we obtain

$$\begin{aligned} \int dq_f P(q_f, t_f \mid q_0, t_0) &= 1 \\ &= J^{-1}(t_f - t_0)^{-1/2} \int_{-\infty}^{\infty} dq_f e^{-\frac{(q_f - q_0)^2}{2D(t_f - t_0)}} \\ &= J^{-1}(t_f - t_0)^{-1/2} \left(2\pi D(t_f - t_0)\right)^{1/2}, \end{aligned} \quad (5.26)$$

implying

$$J = (2\pi D)^{1/2}. \quad (5.27)$$

Finally, for the free case we find

$$P^{free}(q_f, t_f \mid q_0, t_0) = \left(2\pi D(t_f - t_0)\right)^{1/2} \exp\left[-\frac{(q_f - q_0)^2}{2D(t_f - t_0)}\right]. \quad (5.28)$$

that, as could be expected, agrees with the known result for the Wiener process (van Kampen (2004); Gardiner (2009)). Our final result is then

$$P(q_f, t_f \mid q_0, t_0) = \left[2\pi D \left(\frac{\partial q(t_f)}{\partial \dot{q}(t_0)}\right)\right]^{1/2} e^{-S[q_r(t)]/D}. \tag{5.29}$$

In the higher dimensional case, what must be calculated is the van Vleck's determinant; that is the determinant of the matrix with elements $\partial q_j(t_f)/\partial q_k(t_0)$, as was indicated after (5.20) (Schulman (1981)).

5.2 Examples

Here we give a couple of examples in order to show how the results presented in the previous section, particularly (5.29), work. In both, the associated Lagrangian is quadratic in q and $\dot{q}$ and in such cases, as is well known, it is possible to obtain the exact result.

5.2.1 *Ornstein–Uhlenbeck Problem*

In this case, the Langevin equation will be like (2.1), with $g(q,t) = 1$, and $f(q,t) = -bq$. This problem plays within stochastic processes, a similar role as the harmonic oscillator plays in quantum mechanics. The associated FPE is given by

$$\frac{\partial}{\partial t} P(q, t \mid q_0, t_0) = \frac{\partial}{\partial p}(bqP) + \frac{D}{2}\frac{\partial^2}{\partial q^2} P. \tag{5.30}$$

According to (2.18), the corresponding Lagrangian is

$$\mathcal{L}[q(t), \dot{q}(t)] = \frac{1}{2}(\dot{q} + bq)^2 - \alpha b D. \tag{5.31}$$

According to the discretization procedure indicated in Chapter 3, here we have used α. Latter we will see what is the choice adequate to obtain the correct result. The Euler–Lagrange equation is given by

$$\ddot{q} - b^2 q = 0. \tag{5.32}$$

The boundary conditions that the solution must fulfill are ($t_0 = 0, t_f = T$): $q(0) = q_0$ and $q(T) = q_f$. The classical path solution is

$$q_r(t) = \left(\frac{q_f \sinh bt + q_0 \sinh b(T-t)}{\sinh bT}\right). \tag{5.33}$$

From the last equation we obtain

$$\frac{\partial q_r(T)}{\partial \dot{q}_0} = \frac{\sinh bT}{b}. \tag{5.34}$$

On the other hand, the classical action is

$$\mathcal{S}[q_r(t)] = \int_0^T ds\mathcal{L}[q_r(s), \dot{q}_r(s)]$$
$$= b\frac{\left(q_f - q_0 e^{-bT}\right)^2}{(1 - e^{-2bT})} - \alpha bdT. \tag{5.35}$$

Replacing in (5.29), and adopting $\alpha = 1/2$, we finally find

$$P(q_f, t_f \mid q_0, t_0) = \left(\frac{b}{\pi D(1 - e^{-2bT})}\right)^{1/2} \exp\left(-\frac{b\left(q_f - q_0 e^{-bT}\right)^2}{D(1 - e^{-2bT})}\right), \tag{5.36}$$

which completely agrees with the very well known result (van Kampen (2004); Gardiner (2009)).

5.2.2 *Simplified Prey-Predator Model*

Here we present a two dimensional problem that is a simplified version of a Lotka–Volterra or prey-predator model (van Kampen (2004); Wio (1994)),

$$\dot{q}_1(t) = bq_2(t) + \xi_1(t),$$
$$\dot{q}_2(t) = -bq_1(t) + \xi_2(t). \tag{5.37}$$

In order to simplify the calculation we assume that $<\xi_1(t)>=<\xi_2(t)>= 0$, and $<\xi_j(t)\xi_k(t')>= 2D\delta_{jk}\delta(t-t')$. The Lagrangian takes then the form

$$\mathcal{L}[q_1, q_2, \dot{q}_1, \dot{q}_2] = \frac{1}{2}\left\{(\dot{q}_1 - bq_2)^2 + (\dot{q}_2 + bq_1)^2\right\}, \tag{5.38}$$

and the Euler–Lagrange equations are

$$\ddot{q}_1(t) - 2b\dot{q}_2 - b^2 q_1 = 0$$
$$\ddot{q}_2(t) + 2b\dot{q}_1 - b^2 q_2 = 0. \tag{5.39}$$

The classical path, solution of the previous equations, fulfilling at $t = 0, t = T$ the conditions $q_1(0) = q_{1,0}, q_2(0) = q_{2,0}, q_1(T) = q_{1,f}, q_2(T) = q_{2,f}$, is given by

$$q_{1,r}(t) = \frac{t}{T}\left(q_{1,f}\cos(b(T-t)) - q_{2,f}\sin(b(T-t))\right)$$
$$+ \left(1 - \frac{t}{T}\right)\left(q_{1,0}\cos(bt) + q_{2,0}\sin(bt)\right)$$
$$q_{2,r}(t) = \frac{t}{T}\left(q_{1,f}\sin(b(T-t)) + q_{2,f}\cos(b(T-t))\right)$$
$$+ \left(1 - \frac{t}{T}\right)\left(q_{1,0}\sin(bt) - q_{2,0}\cos(bt)\right). \tag{5.40}$$

Hence, the corresponding van Vleck's determinant is equal to T^2.

The classical action, corresponding to the integral of the Lagrangian (5.38) along the trajectory (5.40), becomes

$$\begin{aligned}\mathcal{S}[q_r] = \frac{1}{2T} \Big\{ & q_{1,f}^2 + q_{2,f}^2 + q_{1,0}^2 + q_{2,0}^2 \\ & -2q_{1,f}\left(q_{1,0}\cos(bT) + q_{2,0}\sin(bT)\right) \\ & +2q_{2,f}\left(q_{1,0}\sin(bT) - q_{2,0}\cos(bT)\right)\Big\}. \end{aligned} \tag{5.41}$$

Collecting the results of van Vleck's determinant (T^2) and (5.41), we finally obtain

$$\begin{aligned} P(q_{1,f}, q_{2,f}, T \mid q_{1,0}, q_{2,0}, 0) = {} & (2\pi DT)^{1/2} \\ & \exp\left[-\frac{1}{2DT}(q_{1,f}^2 + q_{2,f}^2 + q_{1,0}^2 + q_{2,0}^2)\right] \\ & \times \exp\left(\frac{1}{2DT}[q_{1,f}(q_{1,0}\cos bT + q_{2,0}\sin bT) \right. \\ & \left. + q_{2,f}(q_{1,0}\sin bT - q_{2,0}\cos bT)]\right) \end{aligned} \tag{5.42}$$

In the limit of null drift ($b \to 0$), according to (5.37), the problem reduces to two Wiener processes without interaction among them. It is easy to check that (5.42) reduces to the indicated result. Also, if we take $D \to 0$, corresponding to the zero noise limit, we recover the propagator for the deterministic problem, in terms of a product of δ-functions, as expected. In this last situation, it becomes clear that this system is also equivalent to a harmonic oscillator in a phase-space representation (q_1: coordinate, q_2: conjugate momentum).

Chapter 6

Space-Time Transformation II

6.1 Introduction

In this chapter we present the exact solution for the diffusion propagator in a time-dependent nonharmonic oscillator $V(x,t) = \frac{1}{2}a(t)x^2 + b\ln x$. This particular choice of the potential could be useful to model the behavior of several physical and biological systems. Among them, the study of neuron models (e.g. integrate and fire models, Burkitt (2006)), stochastic resonance in monostable nonlinear oscillators (Gammaitoni *et al.* (1998)) and its possible application to spatially extended systems (Wio, *et al.* (2002); Wio and Deza (2007)). Also, we can consider that the logarithmic term, within the path integral scheme (or as a Boltzmann like weight) mimics a prefactor corresponding to an effective energy barrier. It is clear that the possibility of having exact expressions for the stochastic propagator in a non-symmetrical potential can be of interest. In fact, in many of the above mentioned applications (specially on neuron models) the potential studied in this work would represent a more realistic approximation to the behavior of the system under study. In other problems such us Brownian motors, this asymmetry is not just an improvement but an unavoidable ingredient of the model.

The approach used here has been inspired by the solution presented in Khandekar and Lawande (1975, 1986); Khandekar *et al.* (2000), corresponding to the exact quantum mechanical propagator in a time-dependent harmonic potential plus a singular perturbation. In the present case, the fact that the metric of the underlying space is Euclidean, allows us to obtain the exact analytical expression of the diffusion propagator for a whole family of functional forms of the time-dependent elastic parameter.

In the next section we introduce the model we are going to study and show the procedure to be followed in order to obtain the exact form of the propagator. We also discuss the presence of noise induced flow of particles through the infinite barrier located at the origin, provided that the noise amplitude is large enough for the particles to overcome the deterministic drift. Afterwards we show how to obtain a family of analytical solutions, and comment on the possible applications of the present results.

6.2 The Diffusion Propagator

In this section we will follow, and adequately adapt, the results in (Khandekar and Lawande (1975, 1986); Khandekar *et al.* (2000)). Our starting point is to consider the following Langevin equation

$$\dot{x} = h(x,t) + \xi(t), \tag{6.1}$$

where $\xi(t)$, as usual, is an additive *Gaussian white noise*. That is, it fulfills the conditions $\langle\xi(t)\rangle = 0$ and $\langle\xi(t)\xi(t')\rangle = 2D\delta(t-t')$. This equation describes the overdamped motion of a particle in a time-dependent potential. Here, we consider the force term $h(x,t) = -a(t)x - b/x$ which, through the relation $h(x,t) = -\partial V/\partial x$, corresponds to the following potential

$$V(x,t) = \frac{1}{2}a(t)x^2 + b\ln x. \tag{6.2}$$

This potential is defined for $x > 0$ and, whenever $b < 0$ and the elastic parameter $a(t)$ is positive, it corresponds to an nonharmonic monostable system composed by a time-dependent harmonic oscillator plus a logarithmic term which is singular at the origin.

As it will be shown later, even in this monostable situation, the noise could be able to induce a flow of particles through the infinite barrier located at the origin, overcoming the deterministic drift whenever $D > -2b$ holds. In fact, the meaningful condition related to the conservation of particles inside the system (zero flux at $x = 0$) is $D < -2b$.

Here we will relax the monostability condition, allowing for the time-dependent elastic parameter to take negative values. We will show that in this extended situation an asymptotic probability distribution can be reached whenever the elastic term satisfies,

$$\lim_{t_0 \to -\infty} \int_{t_0}^{t} a(s)ds = \infty \qquad \forall\, t. \tag{6.3}$$

We will say that in this case the potential is *strongly attractive*.

The path integral representation of $P(x_b, t_b \mid x_a, t_a)$, that is the transition probability associated with this Langevin equation, is given by (Langouche *et al.* (1982))

$$P(x_b, t_b \mid x_a, t_a) = \int_{x(t_a)=x_a}^{x(t_b)=x_b} \mathcal{D}[x(t)] \ \exp\left[-\int_{t_a}^{t_b} L(x(\tau), \dot{x}(\tau), \tau) d\tau\right]. \tag{6.4}$$

Here, in a midpoint discretization, the stochastic *Lagrangian* or *Onsager–Machlup* functional is given by

$$L(x, \dot{x}, t) = \frac{1}{2D}[\dot{x} - h(x,t)]^2 + \frac{1}{2}\frac{\partial h(x,t)}{\partial x}. \tag{6.5}$$

Replacing the actual form of $h(x,t)$ the previous expression can be expanded to yield

$$L = L_0 + \frac{d\Phi}{dt}, \tag{6.6}$$

where Φ corresponds to

$$\Phi(t) = \left[\frac{b}{D} - \frac{1}{2}\right] \int_{t_0}^{t} a(\tau) d\tau + \frac{b}{D} \ln x + \frac{a(t) x^2}{2D}, \tag{6.7}$$

with arbitrary t_0, and

$$L_0 \equiv \frac{1}{2D} \left[\dot{x}^2 + \left(a(t)^2 - \dot{a}(t)\right) x^2 + (b + D)\frac{b}{x^2}\right]. \tag{6.8}$$

Hence the path integral in (6.4) adopts the form

$$P(x_b, t_b | x_a, t_a) = e^{-[\Phi(t_b) - \Phi(t_a)]} K(x_b, t_b | x_a, t_a), \tag{6.9}$$

with

$$K(x_b, t_b \mid x_a, t_a) = \int_{x_a}^{x_b} \mathcal{D}[x(t)] \times \exp\left[-\frac{1}{2D}\int_{t_a}^{t_b} \left(\dot{x}^2 + w(\tau) x^2 + (b + D)\frac{b}{x^2}\right) d\tau\right], \tag{6.10}$$

and $w(t) = a(t)^2 - \dot{a}(t)$. As usual, the path integral in (6.10) is defined in a discrete form by

$$K(x_b, t_b \mid x_a, t_a) = \lim_{N\to\infty} A_N \int \dots \int \exp\left(-\sum_{j=1}^{N} S_j(x_j, x_{j-1})\right) \prod_{j=1}^{N-1} dx_j, \tag{6.11}$$

with $N\varepsilon = t_a - t_b$, $t_j = t_a + j\varepsilon$, $x_0 = x_a$; $x_N = x_b$; $A_N = [2\pi D\varepsilon]^{-\frac{N}{2}}$; and
$S_j = S_j(x_j, x_{j-1}) = \varepsilon\, L_o(x_j, x_{j-1})$

$$= \frac{1}{2D}\left[\frac{x_j^2 + x_{j-1}^2}{\varepsilon} + \varepsilon w_j x^2\right] - \left[\frac{x_j x_{j-1}}{D\varepsilon} + \frac{D(\theta^2 - \frac{1}{4})\varepsilon}{2x_j x_{j-1}}\right].$$

Here

$$\theta = \theta(b) = \frac{1}{2}\sqrt{1 + \frac{4b(b+D)}{D^2}} = \left|\frac{1}{2} + \frac{b}{D}\right|. \tag{6.12}$$

Up to first order in ε (exploiting that $\varepsilon \ll 1$) we can use the following asymptotic form of the modified Bessel function (Watson (1962))

$$\exp\left(\frac{u}{\varepsilon} - \frac{1}{2}\left(\theta(b)^2 - \frac{1}{4}\right)\frac{\varepsilon}{u} + O(\varepsilon^2)\right) \approx \sqrt{\frac{2\pi u}{\varepsilon}}\, I_{\theta(b)}\left(\frac{u}{\varepsilon}\right). \tag{6.13}$$

Using the last expression with $u = x_j\, x_{j-1} \,/\, D$ the propagator of (6.11) may be cast into the following form

$$K(x_b, t_b \mid x_a, t_a) = \lim_{N\to\infty}\left(\frac{1}{2\pi D\varepsilon}\right)^{1/2} \int \cdots \int \prod_{j=1}^{N-1} dx_j$$

$$\times \prod_{j=1}^{N} \exp\left[-\frac{1}{2D\varepsilon}(x_j^2 + x_{j-1}^2 + \varepsilon^2 w_j x_j^2)\right],$$

$$\times \left(\frac{2\pi x_j x_{j-1}}{D\varepsilon}\right)^{1/2} I_\theta\left(\frac{x_j x_{j-1}}{D\varepsilon}\right). \tag{6.14}$$

The last expression can be rewritten as

$$K(x_b, t_b \mid x_a, t_a) = \exp\left[\frac{-\beta}{2}\left(x_a^2 + x_b^2\right)\right] \lim_{N\to\infty} \beta^N \int \cdots$$

$$\cdots \int \prod_{j=1}^{N-1} e^{-\alpha_j\, x_j^2}\, I_\theta\left(\beta x_j x_{j-1}\right) x_j dx_j, \tag{6.15}$$

where,

$$\alpha_j = \beta\left(1 + \frac{\varepsilon^2}{2} w_j\right); \quad 0 \le j \le N-1; \quad \text{and} \quad \beta = \frac{1}{D\varepsilon}.$$

Now, in order to perform the integrations of (6.15), we can use the equality known as Weber formula (Watson (1962)), which is given by

$$\int_0^\infty e^{-\alpha x^2}\, I_\theta(ax)\, I_\theta(bx)\, x\, dx = \frac{1}{2\alpha}\exp\left[\frac{a^2 + b^2}{4\alpha}\right] I_\theta\left(\frac{ab}{2\alpha}\right),$$

and is valid for $\Re(\theta) > -1$, $\Re(\alpha) > 0$ (here both conditions are fulfilled). The final result is

$$K(x_b, t_b \mid x_a, t_a) = \sqrt{x_a x_b} \lim_{N\to\infty} a_N \, e^{\left(p_N x_a^2 + q_N x_b^2\right)} I_\theta \left(a_N x_a x_b\right),$$

where the quantities a_N, p_N and q_N are defined in Appendix A. These quantities are related to a function $Q(t)$ that obeys the equation

$$\ddot{Q}(t) - w(t)Q(t) = 0, \tag{6.16}$$

with the initial condition $Q_0 = Q(t_a) = 0$. In the limit $\varepsilon \to 0$ ($N \to \infty$), we find that (see Appendix A)

$$\lim_{N\to\infty} a_N = \frac{1}{D}\frac{\dot{Q}(t_a)}{Q(t_b)}, \tag{6.17}$$

$$\lim_{N\to\infty} p_N = \lim_{\varepsilon\to 0}\left(\frac{1}{\varepsilon} - \dot{Q}^2(t_a)\int_{t_a+\varepsilon}^{t_b}\frac{dt}{Q(t)^2}\right), \tag{6.18}$$

$$\lim_{N\to\infty} q_N = -\frac{1}{2D}\frac{\dot{Q}(t_b)}{Q(t_b)},. \tag{6.19}$$

To calculate the second limit, it is necessary to solve Eq. (6.16). As it was shown in Batista *et al.* (1996), the complete solution of Eq. (6.16) can be reduced to *quadratures*, with the general form given by

$$Q(t) = k_1\, \mathcal{R}_{t_a}(t) + k_2\, \mathcal{S}_{t_a}(t), \tag{6.20}$$

where

$$\begin{aligned}\mathcal{R}_{t_a}(t) &= e^{-\int_{t_a}^{t} a(s)ds}\\ \mathcal{S}_{t_a}(t) &= e^{-\int_{t_a}^{t} a(s)ds}\int_{t_a}^{t} e^{2\int_{t_a}^{\tau} a(\varsigma)d\varsigma} d\tau.\end{aligned} \tag{6.21}$$

Hence, the solution fulfilling the initial condition $Q_0 = Q(t_a) = 0$ is

$$Q(t) = \dot{Q}(t_a)\, \mathcal{S}_{t_a}(t). \tag{6.22}$$

After replacing this solution into the expressions for a_N, p_N and q_N (see Appendix A), we finally arrive to a completely analytical expression for the transition probability

$$\begin{aligned}P(x_b, t_b \mid x_a, t_a) =& e^{-[\Phi(t_b)-\Phi(t_a)]}\frac{\sqrt{x_a x_b}}{D\,\mathcal{S}_{t_a}(t_b)} \times I_\theta\left(\frac{x_a x_b}{D\,\mathcal{S}_{t_a}(t_b)}\right)\\ &\times \exp\left(\frac{-1}{2D\,\mathcal{S}_{t_a}(t_b)}\left[\left(\mathcal{R}_{t_a}(t_b) + a(t_a)\right)x_a^2 + \left(\frac{1}{\mathcal{R}_{t_a}(t_b)} - a(t_b)\right)x_b^2\right]\right),\end{aligned} \tag{6.23}$$

which can be further simplified as

$$P(x_b, t_b \mid x_a, t_a) = x_a^{\frac{b}{D}+\frac{1}{2}} x_b^{-\frac{b}{D}+\frac{1}{2}} \frac{[\mathcal{R}_{t_a}(t_b)]^{\frac{b}{D}-\frac{1}{2}}}{D\, \mathcal{S}_{t_a}(t_b)} \times I_\theta\left(\frac{x_a x_b}{D\, \mathcal{S}_{t_a}(t_b)}\right)$$
$$\times \exp\left(\frac{-1}{2D\, \mathcal{S}_{t_a}(t_b)}\left[\mathcal{R}_{t_a}(t_b) x_a^2 + \frac{1}{\mathcal{R}_{t_a}(t_b)} x_b^2\right]\right). \tag{6.24}$$

It is straightforward to check that for some particular choices of $a(t)$ and b the last expression fulfills the corresponding Fokker–Planck equation. Albeit not so simple, it is also possible to prove it for the general case. The last expression also indicates that, in order to have the explicit form of the propagator we only need to obtain the function $Q(t)$ (the solution of (6.16) given in (6.20)) for the problem under study (that is, for a given form of the function $a(t)$). We will provide a family of solutions for a rather general form of the function $a(t)$ in a subsequent section. Previously, we will discuss the possibility of finding a net current at the origin.

6.3 Flow Through the Infinite Barrier

Let us first evaluate from (6.24) the asymptotic probability distribution, that is

$$P(x, t) = \lim_{t_a \to -\infty} P(x, t \mid x_a, t_a). \tag{6.25}$$

In the strongly attractive case (i.e. (6.3)), it can be easily shown that $\mathcal{S}_{t_a}(t)$ diverges and that $\mathcal{R}_{t_a}(t)$ goes to zero as $t_a \to -\infty$. Thus, we will make use of the expansion of the modified Bessel function for small argument (Watson (1962)),

$$I_\theta(z) = \frac{1}{\Gamma(\theta+1)}\left(\frac{z}{2}\right)^\theta + \mathcal{O}(z^{\theta+2}). \tag{6.26}$$

Replacing this expansion into (6.24) we get,

$$P(x, t) = \left[\lim_{t_a \to -\infty} [x_a\, \mathcal{R}_{t_a}(t)]^{\frac{b}{D}+\frac{1}{2}+\theta}\right] \frac{2}{\Gamma(\theta+1)}$$
$$\times \frac{x^{-\frac{b}{D}+\frac{1}{2}+\theta}}{(2\, D\, g(t))^{1+\theta}} \exp\left(-\frac{x^2}{2\, D\, g(t)}\right), \tag{6.27}$$

where $g(t)$ is defined as

$$g(t) = \lim_{t_a \to -\infty} \mathcal{S}_{t_a}(t)\mathcal{R}_{t_a}(t). \tag{6.28}$$

It is clear that unless the condition

$$1/2 + b/D + \theta = 0, \tag{6.29}$$

holds, the system cannot reach an asymptotic probability distribution. In fact, the term between the square brackets in (6.27) depends on the initial condition. Furthermore, it can be shown that the normalization of (6.27) gives a vanishing function of t unless the previous condition holds. Note that (6.29) implies, through (6.12), $D < -2b$. This condition gives the maximum value of noise amplitude for the particles to be confined inside the interval $(0, \infty]$.

We want now to make explicit the existence of a noise induced probability current through the infinite barrier at the origin when $D > -2b$. Before giving a rigorous deduction, let us state a simple argument which provides some clue about the underlying physical mechanism governing this flow. As it is clear from the Langevin equation, the particle is subjected to both deterministic and stochastic forces. If we analyze separately both contributions to the particle movement near the origin, we obtain for the deterministic trajectory $x_d(t) = \sqrt{-2bt}$. Comparing this result with the well known diffusive behavior, where the uncertainty on the particle's position grows as $x_s = \sqrt{Dt}$, we reobtain the previous condition $D > -2b$ for the possible appearance of *noise induced leakage* of particles.

The probability current at the origin $J(x = 0, t \mid x_a, t_a)$ can be evaluated from the associated Fokker–Planck equation. In the case $D > -2b$ we obtain

$$\begin{aligned} J(x = 0, t \mid x_a, t_a) &= \left(-ax - \frac{b}{x} - \frac{D}{2}\frac{\partial}{\partial x} \right) P(x, t \mid x_a, t_a) \Big|_{x=0} \\ &= -D\left(\frac{b}{D} + \frac{1}{2} \right) J_{x_a,t_a}(t), \end{aligned} \tag{6.30}$$

where it can be shown that $J_{x_a,t_a}(t)$ is a positive function of time for any given initial condition. Therefore, we have obtained a non-zero negative current, as previously stated[1].

[1]This somewhat counter-intuitive flux has an interesting quantum counterpart in the *fall to the center* effect studied in Sec. 35 of Landau and Lifshitz (1958).

6.4 Asymptotic Probability Distribution

In the case $D < -2b$ there is no probability leakage. In fact, the asymptotic probability distribution can be obtained from (6.31) and is given by

$$P(x,t)= \frac{2}{\Gamma(\theta+1)} \frac{x^{|\frac{2b}{D}|}}{(2\,D\,g(t))^{1+\theta}} \exp\left(-\frac{x^2}{2\,D\,g(t)}\right), \qquad (6.31)$$

which can be easily shown to be normalizable.

It is worth to study how the properties of the elastic parameter function $a(t)$ influences the behavior of the function $g(t)$, which reflects the time evolution of the width of the probability distribution. First, note that from the definition of $g(t)$ given in (6.28) we can obtain

$$\dot{g}(t) = -2a(t)g(t) + 1. \qquad (6.32)$$

From this equation it can be deduced that $g(t) > 0\ \forall t$, as must be expected for any well behaved probability distribution. In addition, it can be proved that in order to confine the particle in a small region of width $\sqrt{g(t)} \sim \sqrt{\epsilon}$ an attractive force of order $a(t) \sim 1/\epsilon$ is needed. On the other hand, a small attractive force of order $a(t) \sim \epsilon$, gives a broad distribution with $g(t) \sim 1/\epsilon$. The limiting cases for $P(x,t)$ corresponding to an unbounded spreading ($a(t) \to 0 \Rightarrow g(t) \to \infty$), and to an asymptotically approach to a $\delta(x)$ distribution ($a(t) \to \infty \Rightarrow g(t) \to 0$), can be also obtained.

From the previous paragraph it is clear that even in the strongly attractive situation (see (6.3)) the probability distribution may exhibit an unbounded spreading. In fact, we have already shown that even in the monostable situation $a(t) > 0$, but where the strength vanishes in time, $g(t)$ grows indefinitely. Therefore, in order to obtain a non-divergent width of the probability distribution, the conditions on the attractive term have to be stronger than the one imposed by (6.3). We may infer that the localized-probability condition should be related to a non-vanishing attractive strength of the time averaged potential.

6.5 General Localization Conditions

Let us discuss the set of conditions that ensures the asymptotic localization of the probability distribution. From the analysis of (6.32) it is clear that in order to guarantee a non divergent $g(t)$ the elastic parameter should have the following properties. Its accumulated strength is positive, i. e.

$$\int_{t_i}^{t} a(\tau)\, d\tau = c > 0 \qquad \forall t, \qquad (6.33)$$

where c is an arbitrary constant and t_i is the nearest time which fulfills the previous equation. This condition is clearly fulfilled if the potential is strongly attractive. We may also infer that the accumulated attractive effect (see (6.33)) should be non-vanishing. In other words, the elapsed time where the accumulated strength reaches the given constant c is bounded, that is

$$t - t_i = \Delta t \leq \Delta t_u \qquad \forall t, \tag{6.34}$$

where $\Delta t_u \equiv \Delta t_u(c)$ is the mentioned upper bound for the elapsed time. It is evident that condition (6.34) is more restrictive than the one imposed by (6.3).

In the following section we will provide a family of examples where the probability is asymptotically localized.

6.6 A Family of Analytical Solutions

As already mentioned, to obtain the final expression for the diffusion propagator, we must first solve (6.16) for a given choice of the function $a(t)$. Because the frequency $\omega(t)$ depends only on the harmonic term of the potential, we can make use of any known solution of the simpler harmonic case. A method to generate a whole family of analytical solutions has been proposed in Batista *et al.* (1996) for the time-dependent harmonic oscillator. In order to reach such a goal the elastic parameter was written in the following form

$$a(t) = f(t) + \frac{1}{2}\frac{\dot{f}(t)}{f(t)} \tag{6.35}$$

This allows us to find the corresponding independent solutions of (6.16)

$$q_1(t) = \frac{\sinh(F(t))}{\sqrt{f(t)}}, \tag{6.36}$$

$$q_2(t) = \frac{\cosh(F(t))}{\sqrt{f(t)}}, \tag{6.37}$$

where $F(t) = \int_{t_0}^{t} f(s)\, ds$, indicating that $f(t)$ must be an integrable function. The solution which satisfies the initial condition $Q(t_a) = 0$ reduces to

$$Q(t) = \dot{Q}(t_a)\frac{\sinh(F(t) - F(t_a))}{\sqrt{f(t)f(t_a)}}. \tag{6.38}$$

With this result, the transition probability in (6.24) adopts the analytical form

$$
\begin{aligned}
P(x_b, t_b \mid x_a, t_a) &= \mathrm{e}^{\left[\frac{1}{2} - \frac{b}{D}\right](F(t_b) - F(t_a))} \left(\frac{f(t_a)}{f(t_b)}\right)^{\frac{b}{2D} - \frac{1}{4}} x_a^{\frac{b}{D} + \frac{1}{2}} x_b^{-\frac{b}{D} + \frac{1}{2}} \\
&\quad \times \frac{\sqrt{f(t_b) f(t_a)}}{D \sinh(F(t_b) - F(t_a))} I_{\theta(b)} \left(\frac{x_a x_b}{D} \frac{\sqrt{f(t_b) f(t_a)}}{\sinh(F(t_b) - F(t_a))}\right) \\
&\quad \times \exp\left(-\frac{f(t_a) e^{-(F(t_b) - F(t_a))} x_a^2 + f(t_b) e^{(F(t_b) - F(t_a))} x_b^2}{2D \sinh(F(t_b) - F(t_a))}\right).
\end{aligned}
\tag{6.39}
$$

Hence, we have obtained a completely analytical expression for the propagator in (6.39), that only depends on the choice of the elastic parameter $a(t)$. A whole family of solutions are shown in the Appendices in Batista *et al.* (1996).

It is worth to remark here that the limit $b \to 0$ is a (kind of) singular one. The naive point of view will be that, in such a limit, the form of the propagator in (6.24) shall reduce to the one corresponding to the case of the harmonic time-dependent potential $V(x) \sim \frac{1}{2} a(t) x^2$. However, this limit corresponds to a harmonic time-dependent potential for $x > 0$ with an absorbing boundary condition at $x = 0$. Then, it should be possible to re-obtain the limit $b \to 0$ of the diffusion propagator found in this paper ($P_0(x_b, t_b \mid x_a, t_a)$) from the one obtained in Batista *et al.* (1996) for the harmonic time-dependent case ($P_h(x_b, t_b \mid x_a, t_a)$) simply as

$$
P_0(x_b, t_b \mid x_a, t_a) = P_h(x_b, t_b \mid x_a, t_a) - P_h(-x_b, t_b \mid x_a, t_a).
$$

It can be easily proved that this is indeed the case.

6.7 Stochastic Resonance in a Monostable Non-Harmonic Time-Dependent Potential

The knowledge of the exact form of the above found propagator can be useful to model different physical and biological phenomena. Particularly interesting problems, suitable to be studied taking advantage of the above results, are realistic non-symmetric neuron membrane potentials and the phenomenon of stochastic resonance (SR) in a monostable zero-dimensional potential, in spatially extended systems, and in several neuron firing models (Burkitt (2006)), among others. A complete review of these SR topics can be found in Gammaitoni *et al.* (1998). The knowledge of the exact propagator in the indicated time-dependent nonharmonic potential can be useful as a benchmark to test approximate numerical or analytical procedures.

Among the several studies of SR in monostable systems, it has been shown using scaling arguments and numerical experiments, that the signal-to-noise ratio (SNR) is a monotonically increasing function of the noise amplitude. By contrast, it is quite clear that this increase in the response of the system cannot be unbounded. Let us now exploit the previous results in order to study SR in a monostable potential. There are a few recent papers where this situation has been studied by means of linear response theory and analogical simulation in a single-well Duffing oscillator, as well as by scaling analysis and numerical simulations. These findings conform a non-conventional form of SR (see Gammaitoni *et al.* (1998) and references therein).

Here we present a completely analytical analysis of SR in a monostable non-harmonic potential. As indicated, we exploit the previously obtained form of the exact solution for the diffusion propagator in the time-dependent non-harmonic oscillator $V(x,t) = \frac{a(t)}{2}x^2 + b\ln x$ (that is, the same potential analyzed before).

Our starting point is the Langevin equation

$$\dot{x} = h(x,t) + \xi(t), \tag{6.40}$$

where $\xi(t)$ is an additive *Gaussian white noise.* This equation describes the overdamped motion of a particle in a time-dependent potential. Here we will consider a force term $h(x,t) = -a(t)x - b/x$ corresponding to the potential

$$V(x,t) = \frac{1}{2}a(t)x^2 + b\ln x. \tag{6.41}$$

As we have seen, this potential is defined for $x > 0$, and whenever $b < 0$ corresponds to a non-harmonic monostable system composed by a time-dependent harmonic oscillator plus a logarithmic term which is singular at the origin. We have also seen that the exact form of the propagator, written in a symmetric and compact form, is given by

$$P(x_b, t_b|x_a, t_a) = AB^{-(1+\theta)} x_a^{-\theta} x_b^{(1+\theta)} I_\theta(Ax_a x_b)$$
$$\exp\left(-\frac{A}{2}(Bx_a^2 + \frac{1}{B}x_b^2)\right). \tag{6.42}$$

Here we have defined

$$A = A(t_a, t_b) = \left[D e^{-\int_{t_a}^{t_b} a(s)ds} \int_{t_a}^{t_b} e^{2\int_{t_a}^{\tau} a(s')ds'} d\tau\right]^{-1}, \tag{6.43}$$

and

$$B = B(t_a, t_b) = e^{-\int_{t_a}^{t_b} a(s)ds}, \tag{6.44}$$

and the parameter θ is given by $\theta = -\left(\frac{1}{2} + \frac{b}{D}\right) > 0$, which is always positive under the zero flux condition at $x = 0$.

The one-time probability distribution, that can be obtained from (6.42) in the limit $t_a \to -\infty$, is

$$P(x, t) = \lim_{t_a \to -\infty} P(x, t|x_a, t_a)$$
$$= \frac{2}{\Gamma(\theta + 1)} \left(\frac{g(t)}{2\,D}\right)^{1+\theta} x^{1+2\theta} \exp\left(-\frac{g(t)x^2}{2D}\right), \tag{6.45}$$

with

$$g(t) = \lim_{t_a \to -\infty} \left(\frac{A(t_a, t)D}{B(t_a, t)}\right)$$
$$= \lim_{t_a \to -\infty} \left\{e^{2\int_{t_a}^{t} a(s)ds} \left(\int_{t_a}^{t} e^{2\int_{t_a}^{\tau} a(s')ds'} d\tau\right)^{-1}\right\}. \tag{6.46}$$

To obtain the asymptotic one-time distribution as written in (6.45), it is necessary that the long time behavior of the potential results to be attractive, i.e. $\lim_{t' \to -\infty} B(t', t) = 0$; $\lim_{t' \to -\infty} A(t', t) = 0$; The knowledge of the probability distribution given in (6.45) allows us to analytically evaluate its moments.

It is worth to point out that in the weak noise limit ($D \to 0$), these results are consistent with those arising from the evaluation of the deterministic trajectory. Such a trajectory is easy to obtain and is given by $x_{det}(t) = \sqrt{\frac{-2b}{g(t)}}$.

From both, (6.42) and (6.45), we can obtain the two-time joint probability and using it we arrive to the following expression for the autocorrelation function

$$\langle x(t_a)\; x(t_b)\rangle = \frac{2D}{\sqrt{g(t_a)g(t_b)}} \frac{\Gamma^2(\theta + \frac{3}{2})}{\Gamma^2(\theta + 1)} \mathcal{F}\left(-\frac{1}{2}, -\frac{1}{2}; \theta + 1; z\right). \tag{6.47}$$

where

$$z = \frac{\int_{-\infty}^{t_a} e^{2\int_{-\infty}^{\tau} a(\xi)d\xi} d\tau}{\int_{-\infty}^{t_b} e^{2\int_{-\infty}^{\tau} a(\xi)d\xi} d\tau}, \tag{6.48}$$

and $\mathcal{F}$ stands for the hypergeometric function (Abramowitz and Stegun (1964)).

In order to study the phenomenon of SR in the monostable nonharmonic potential $V(x) = ax^2 + b\ln x$, we will make use of the previous results to analyze the effects of a small harmonic time–dependent perturbation on the correlation function. We notice that a very simple form $g(t) = 2a_o + 2\alpha\sin(\omega t)$ ($\alpha \ll 1$) yields, solving the integral equation (6.46), the desired time dependence of $V(x,t)$, up to first order in the perturbation parameter α, i.e.

$$V(x,t) = a_o x^2 (1 + \frac{\alpha}{K}\sin(\omega t + \phi)) + b\ln x + \mathcal{O}(\alpha), \tag{6.49}$$

where $\phi \simeq \omega/2a_o \ll 1$ (that corresponds to the case where the phase shift between the potential minimum and the deterministic trajectory is small).

Expanding the hypergeometric function $\mathcal{F}$ up to second order in α we can obtain an approximate expression for the correlation function. From it, defining

$$z_0 = z|_{\alpha=0} = e^{-2K\tau} \tag{6.50}$$

$$\bar{D} = 2D\frac{\Gamma^2(\theta+\frac{3}{2})}{\Gamma^2(\theta+1)} \tag{6.51}$$

$$\mathcal{F}_\theta^n = \mathcal{F}\left(-\frac{1}{2}+n, -\frac{1}{2}+n; \theta+1+n; z_0\right), \tag{6.52}$$

and also expanding the product $g(t_a)g(t_b)$, and making a time average we obtain for the autocorrelation function

$$\begin{aligned} C(\tau) = \langle x(t)x(t+\tau)\rangle_t &= \frac{\omega}{2\pi}\int_0^{2\pi/\omega} \langle x(t)x(t+\tau)\rangle dt \\ &= \frac{\bar{D}}{K}\mathcal{F}_\theta^0 + \alpha^2\bar{D}\left\{\frac{3+\cos(\omega\tau)}{8K^3}\mathcal{F}_\theta^0 + \frac{z_0\sin(\omega\tau)}{4\omega K^2(\theta+1)}\mathcal{F}_\theta^1\right. \\ &\quad \left. + \frac{1-\cos(\omega\tau)}{K\omega^2}\left[\frac{z_0}{2(\theta+1)}\mathcal{F}_\theta^1 + \frac{z_0^2}{8(\theta+1)(\theta+2)}\mathcal{F}_\theta^2\right]\right\}. \end{aligned} \tag{6.53}$$

Here it is worth remarking that $\bar{D} \neq 0$ when $D \to 0$. Due to the symmetry of $C(\tau)$ on τ, we make the cosine Fourier transform obtaining the power spectral density (PSD) $S(\Omega)$.

Such a $S(\Omega)$ is composed of several contributions. Some of them are worth to be pointed out. Considering the zero noise situation ($D = 0$) we can identify the deterministic contribution to the correlation function. Such a contribution is given by

$$C_{det}(\tau) = \langle x_{det}(t)\; x_{det}(t+\tau)\rangle_t, \tag{6.54}$$

where the deterministic trajectory $x_{det}(t)$ is known. The form of this contribution in the correlation function and the PSD are, respectively

$$C_{det}(\tau) \approx -\frac{b}{a_o} - \frac{b\alpha^2}{8a_o^3}(3 + \cos(\omega\tau)) \tag{6.55}$$

$$S_{det}(\Omega) \approx \left(-\frac{b}{a_o} - \frac{3b\alpha^2}{8a_o^3}\right) 2\pi\delta(\Omega) - \frac{b\alpha^2}{8a_o^3} 2\pi\delta(\Omega - \omega). \tag{6.56}$$

Equation (6.55) can also be obtained from (6.53) in the limit $D \to 0$.

It is worth here remarking a couple of points. Firstly, the mean value of the position is different from zero, leading to the δ contribution in $\Omega = 0$. Secondly, even without noise we find a signal as a result of the oscillation of the potential minimum. The δ contribution located at $\Omega = \omega$ is due to this fact. In order to study the correlations that arise only in presence of noise, we will subtract in the following analysis both deterministic contributions.

We also remark that in presence of noise ($D \neq 0$), and after subtracting the deterministic part, another δ contribution remains at $\Omega = 0$. The origin of this contribution is the shift in the mean particle's position, due to the asymmetry of the potential. Indeed this shift can be calculated and is given by $\langle x(t)\rangle \approx \sqrt{\frac{-2b}{g(t)}}(1 + \frac{1}{8\theta})$. Similar contributions have been observed by several authors.

Now, let us look at the PSD in $\Omega = \omega$, for the case $a_o \gg 1$ and $\omega \gg 1$ (we can always reach such a condition with an adequate change of the time scale). The Fourier transform of the correlation, expanded up to second order in z_0, reads

$$\begin{aligned}\Big[S(\Omega) - S_{det}(\Omega)\Big]_{\Omega=\omega} &= \pi(\bar{D} + b)\frac{\alpha^2}{8a_o^3}\delta(\Omega - \omega) \\ &\quad + \frac{\bar{D}}{4(\theta+1)a_o^2}\left(1 + \frac{1}{16(\theta+2)}\right),\end{aligned} \tag{6.57}$$

that yields for $\mathcal{R}$, the signal-to-noise ratio (SNR)

$$\mathcal{R} = \alpha^2 \frac{\pi(\theta+1)(\bar{D} + b)}{2a_o\bar{D}(1 + \frac{1}{16(\theta+2)})}. \tag{6.58}$$

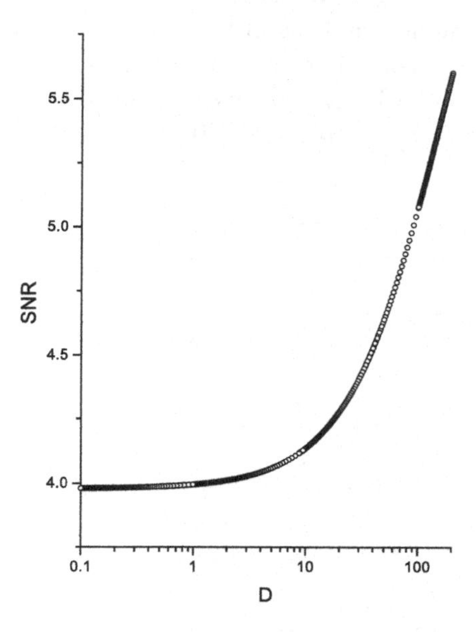

Fig. 6.1 Stochastic resonance response, that is the signal-to-noise ratio, for this nonlinear monostable potential.

The limits of the SNR when $D \to 0$ and $D \to -2b$ (the maximum allowed value of D preventing a probability leaking through the barrier at the origin) are, respectively

$$\mathcal{R}_0 = \alpha^2 \frac{\pi}{8a_o} \tag{6.59}$$

$$\mathcal{R}_{-2b} = \alpha^2 \frac{(\pi - 2)}{2a_o(1 + \frac{1}{32})}. \tag{6.60}$$

These results indicates that the maximum gain becomes independent of all the system parameters $\mathcal{R}_{-2b}/\mathcal{R}_0 \simeq 1.41$.

The analysis of (6.57) that gives the dependence of the SNR on D shows the monotonic enhancement of SNR with the noise intensity. This is in agreement with previous findings. Figure 6.1 shows this result. The present results offer a new ingredient to the previous related ones as we found here, for a (in principle) infinitely bounded potential, a natural cut-off to the SNR enhancement in non-harmonic monostable potentials. The present exact result, in addition to be useful as a benchmark for numerical procedures to be applied to the analysis of more elaborate models, can be adequate to model some problems in biology and chemistry (Gammaitoni *et al.* (1998)). Even more, it can be useful to analyze the effect of the asymmetry of the potential on the response of a neural system modeled through an integrate-an-fire approach (Burkitt (2006)).

Chapter 7

Non-Markov Processes: Colored Noise Case

7.1 Introduction

As we point out at the beginning of Chapter 2, the study of dynamical systems perturbed by noise is recurrent in many contexts of physics and other sciences. Particularly in the theory of non-equilibrium systems, where the macrovariables obey nonlinear equations of motion, noise plays an extremely important role. In fact, the system can overcome potential barriers and reach different macrostates only due to the presence of the noise.

In particular, the consideration of noise sources with finite correlation time (i. e. colored noise), has become a subject of current study in the context of more realistic models of physical systems. For example, in describing the static and dynamical properties of dye lasers it is usual to model the phenomena in terms of stochastic equations where, besides the standard internal white noise, the system is driven by an external colored noise. In a different context, the effect of a non-white spectra in the fluctuations has also been considered in models of gene selection, and in the dynamics of *molecular motors* (Jülicher *et al.* (1997)). These lines of enquiry, and the lack of analytical results, have made more acute the need of analyzing the colored-noise problems from new points of view. Some papers and reviews (Fox (1986); Lindenberg *et al.* (1989); Sancho and San Miguel (1989); Colet *et al.* (1989); Hänggi (1989); Hänggi *et al.* (1990); Luciani and Verga (1988)), offer a view of the state of the art.

Some authors have focused their efforts on the obtention of Markovian approximations, trying to capture the essential features of the original non-Markovian problem. One particular case is the *Unified Colored Noise Approximation* (UCNA), of Hänggi and collaborators (Jung and Hänggi (1987, 1989); Hänggi and Jüng (1995)). The aim of this approximation can be understood in the following way. The original formulation of the problem is in terms of a non-Markovian stochastic differential equation in the relevant variable. However, this problem can be transformed into a Markovian one by extending the number of variables (and equations). The UCNA consists of an adiabatic-like elimination procedure (van Kampen (1985)), that allows to reduce this extended problem to an *effective* Markovian one in the original variable space. The ultimate goal of these procedures is getting a consistent single variable Fokker–Planck approximation for the probability distribution in the original variable.

The UCNA approximation has been justified as a reliable Markovian approximation by means of path integral techniques (Colet *et al.* (1989); Wio *et al.* (1989)). It is such a path integral approach what we will discuss here.

7.2 Ornstein–Uhlenbeck Case

We consider the stochastic process characterized by the following Langevin equation

$$\dot{q} = f(q) + g(q)\eta(t), \tag{7.1}$$

where $\eta(t)$ is an Ornstein–Uhlenbeck process (Gardiner (2009); van Kampen (2004); Wio (1994); Wio, Deza and López (2012)). That is, a Gaussian processes with zero mean and correlation

$$C(t,t') \equiv \langle \eta(t)\eta(t') \rangle = \frac{D}{\tau} \exp\left\{ -\frac{|t-t'|}{\tau} \right\}, \tag{7.2}$$

with D the noise intensity, and τ its correlation time. Equation (7.1), with given initials conditions, defines a non-Markovian process in the configuration space of the variable $q(t)$. In the limit $\tau \to 0$ we have that $\eta(t)$ becomes a white noise with correlation $C(t,t') \to 2D\delta(t-t')$, that is $q(t)$ becomes a Markov process.

As it is well known, the process $q(t)$ might be described by Markovian methods at the price of enlarging the space of variables. If we consider that $\eta(t)$ is an additional variable driven by

$$\dot{\eta} = -\frac{1}{\tau}\eta + \frac{1}{\tau}\xi(t), \tag{7.3}$$

where $\xi(t)$ is a Gaussian white noise with $\langle \xi(t)\xi(t') \rangle = 2D\delta(t-t')$. The set of equations (7.1) and (7.3) define a Markovian process in the enlarged space of variable (q, η). It is characterized by the Fokker–Planck equation

$$\frac{\partial}{\partial t}P(q,\eta,t|q_o,\eta_o,t_o) = -\frac{\partial}{\partial q}\left([f(q)+g(q)\eta]P\right) + \frac{\partial}{\partial \eta}\left(\frac{\eta}{\tau}P\right) + \frac{D}{\tau^2}\frac{\partial^2}{\partial \eta^2}P. \tag{7.4}$$

We are interested in a path integral representation of the process $q(t)$, in the q-configuration space of interest. The derivation of such a representation has two possible routes: one corresponds to start from the original SDE (7.1), the other to work with the enlarged Markov description ((7.1) and (7.3) together). However, as was shown in (Colet *et al.* (1989); Wio *et al.* (1989)), the natural way to treat the delicate point of initial conditions and preparation effects is by means of the enlarged Markov description.

The first route is the one exploited by Pesquera *et al.* (1983), where attention was paid to the trajectory in q space, following the same line of thought for white (DeDominicis and Peliti (1978)) than for colored noise (Phytian (1977, 1980), see also Fox (1986)). Let us sketch such derivation for the simple $g = 0$ case.

Considering the characteristic functional of η we get the probability density of a noise realization as

$$\rho(\eta(s), t_o < s < t) = N\exp\left\{-\frac{1}{2}\int_{t_o}^{t} du \int_{t_o}^{t} du'\eta(u)R(u,u')\eta(u')\right\}, \tag{7.5}$$

where $R(u, u')$ is the inverse of the correlation function $C(u, u')$. The transition probability for the non Markovian process $q(t)$ is obtained considering the SDE (7.1) and (7.5), integrating over all paths $q(t)$ starting at (q_o, t_o) and reaching (q, t). Using the *prepoint* discretization we obtain

$$P(q,\eta,t|q_o,\eta_o,t_o) = \int_{q_o}^{q}\mathcal{D}[q(s)]$$

$$\times\exp\left\{-\frac{1}{2}\int_{t_o}^{t} du \int_{t_o}^{t} du'\left[\dot{q}(u)-f(q(u))\right]R(u,u')\left[\dot{q}(u')-f(q(u'))\right]\right\}, \tag{7.6}$$

Clearly, we have assumed that η is in its stationary state.

We now require the knowledge of $R(u, u')$. Let us call

$$\bar{C}(s,s') = g(q(s))C(s,s')g(q(s')),$$

and we look for its inverse. It is defined according to

$$\int_{t_o}^{t} ds'\bar{C}(s,s')\,\bar{R}(s'-s'') = \delta(s-s''). \tag{7.7}$$

We recall that $C(s,s')$ admits the following formal expansion (when $s > s'$)

$$C(s,s') = \frac{D}{\tau}\exp\left(-\frac{|s-s'|}{\tau}\right)$$

$$= 2D\sum_{n=0}^{\infty}\tau^n\delta^{(n)}(s-s'), \tag{7.8}$$

where $\delta^{(n)}(s-s')$ indicates the derivative of the δ function respect to s'. We have used the convention $\int_0^\infty ds\delta(s) = \frac{1}{2}$. Considering the above indicated form it is natural to propose

$$\bar{R}(s',s'') = \frac{1}{g(q(s'))g(q(s''))}\sum_{n=0}^{\infty}a_n\delta^{(n)}(s'-s''). \tag{7.9}$$

when $t_o < s < t$, splitting the integral in (7.7) for $s' < s$ and $s' > s$ and substituting the above expansions it is easy to get, by simply coefficient identification that (7.9) fulfills (7.7) with $a_n = 0$ for all $n > 2$, and $a_0 = 1/2D$, $a_2 = -\tau^2/2D$. We have the remarkable result that although $\bar{C}(s,s')$ has contributions to all order in τ, the series expansion of its inverse is cut at order τ^2. When $s = t$ or $s = t_o$, the derivation should take into account *surface terms*. We refer to Wio *et al.* (1989) for this aspect.

Now, using the above expression for $R(u,u')$, and after some integration by parts we obtain (but now assuming the general case $g(q) \neq 1$)

$$P(q,\eta,t|q_o,\eta_o,t_o) = \int_{q_o}^{q}\mathcal{D}_g[q(s)]\exp\left\{-\int_{t_o}^{t}du\mathcal{L} - \tau\frac{[\dot{q}_o - f(q_o)]^2}{2Dg(q_o)^2}\right\}, \tag{7.10}$$

where $\mathcal{D}_g$ is a shorthand notation for $\mathcal{D}[q(s)/\int g(q(s))dq(s)]$. The integral is over all path going from (q_o,t_o) to (q,t). Here $\mathcal{L}$ is the Lagrangian-like function

$$\mathcal{L} = \frac{1}{4Dg(q)^2}\left\{\tau\left(\ddot{q} - f'(q)\dot{q} - \frac{g'(q)}{g(q)}\dot{q}[\dot{q} - f(q)]\right) + \dot{q} - f(q)\right\}^2, \tag{7.11}$$

that in the limit $\tau \to 0$ reduces to the well known Lagrangian for the white noise case (in the prepoint discretization). As the Lagrangian (7.11) depends on $\ddot{q}$, reflecting its non-Markovian character, it is not a proper Lagrangian function. A remarkable feature is that, while most studies in colored noise are based in expansions involving power series in τ with infinite order, it only contains linear and quadratic terms in τ.

The second route, exploiting the enlarged Markovian process, uses a standard representation (Langouche *et al.* (1982); DeDominicis and Peliti (1978)) for a n-variable diffusion process. However, the delicate point here is that the associated diffusion matrix is singular. This prevents the direct use of a Lagrangian formulation and requires a phase-space representation where, as was shown in Chapter 2, we need to introduce $(\hat{q}, \hat{\eta})$, that is the *conjugate* momenta for the coordinates (q, η) (Martin, *et al.* (1973)).

The transition probability $P(q, \eta, t|q_o, \eta_o, t_o)$ have the following representation in this enlarged phase-space

$$P(q, \eta, t|q_o, \eta_o, t_o) = \int_{q_o, \eta_o}^{q, \eta} \mathcal{D}[q(s)]\mathcal{D}[\eta(s)]\mathcal{D}[\hat{q}(s)]\mathcal{D}[\hat{\eta}(s)] \exp(\mathcal{S}), \tag{7.12}$$

where the stochastic action is

$$\mathcal{S} = \int_{t_o}^{t} ds \left[i\dot{q}(s)\hat{q}(s) + i\dot{\eta}(s)\hat{\eta}(s) - \mathcal{H}(q(s), \hat{q}(s), \eta(s)\hat{\eta}(s)) \right], \tag{7.13}$$

and the Hamiltonian-like function $\mathcal{H}$ is

$$\mathcal{H} = i\hat{q}(s)\left(f(q) + g(q)\eta\right) - i\hat{\eta}\frac{\eta}{\tau} - \frac{D}{\tau^2}\left(i\hat{\eta}\right)^2. \tag{7.14}$$

It is worth noting that the Lagrangian function could be obtained through a Legendre transformation of the complex function in (7.14) in which $i\hat{q} \to \dot{q}$ and $i\hat{\eta} \to \dot{\eta}$. Hence, we have

$$\mathcal{L}(q, \dot{q}, \eta, \dot{\eta}) = i\dot{q}\hat{q} + i\dot{\eta}\hat{\eta} - \mathcal{H}(q, \eta, \hat{q}, \hat{\eta})$$

$$= \frac{\tau^2}{4D}\left(\dot{\eta} + \frac{\eta}{\tau}\right)^2. \tag{7.15}$$

In the present two variable approach fluctuations are associated to the variable η, while q obeys a deterministic equation. This is the reason why a Lagrangian that weights different possible paths is independent of q. In order to recover the Lagrangian in (7.11) one has to integrate in (7.12) over η, $\hat{\eta}$ and $\hat{q}$. This can be done in various different ways depending on the integration order. Two of these possibilities (corresponding to integration in the following order: $\eta, \hat{\eta}, \hat{q}$ and $\hat{\eta}, \eta, \hat{q}$) require the inversion of the noise correlation function. Such a procedure is particularly interesting because makes clear the reason why $\mathcal{L}$ does not contain terms of the form τ^n, with $n > 2$.

Let us first integrate over the noise variable η and afterwards over its conjugate momentum $\hat{\eta}$. What we obtain is

$$P(q,\eta,t|q_o,\eta_o,t_o) = \left(\frac{4\pi}{\tau}\sinh^{-1}\left[\frac{t-t_o}{\tau}\right]\right)^{-1/2}\int_{q_o,\eta_o}^{q,\eta}\mathcal{D}[q(s)]\mathcal{D}[\hat{q}(s)]$$

$$\times\exp\left(-\mathcal{S}_1 + i\int_{t_o}^{t} ds\hat{q}(s)(\dot{q}-f(q))\right), \quad (7.16)$$

where

$$\mathcal{S}_1 = \left(\frac{4\pi}{\tau}\sinh^{-1}\left[\frac{t-t_o}{\tau}\right]\right)^{-1}\Big\{\eta^2\exp[t/\tau] + \eta_o^2\exp[-t_o/\tau] - 2\eta\eta_o$$

$$-\frac{4D}{\tau}\eta\int_{t_o}^{t} ds\, i\,\hat{q}(s)g(q(s))\sinh\left[\frac{s-t_o}{\tau}\right]$$

$$-\frac{4D}{\tau}\eta_o\int_{t_o}^{t} ds\, i\,\hat{q}(s)g(q(s))\sinh\left[\frac{t-s}{\tau}\right]$$

$$+\frac{8D^2}{\tau^2}\int_{t_o}^{t} ds\int_{t_o}^{t} ds'\hat{q}(s)g(q(s))\sinh\left[\frac{t-s}{\tau}\right]$$

$$\times\sinh\left[\frac{s'-t_o}{\tau}\right]\hat{q}(s')g(q(s'))\Big\}. \quad (7.17)$$

If we integrate over all possible final values of η we get an equation similar to (7.16) and (7.17). If in addition we take into account preparation effects, we should afterwards integrate over a joint distribution of q_o and η_o. The integration over $\hat{q}$ requires the inversion of a time-dependent function. It corresponds to the non-stationary noise correlation function.

Let us focus on the transition probability in q space for the case of stationary noise. It results from (7.16) integrating over all possible final values of η, and over the initial values of η_o associated to a stationary distribution given by

$$P_{st}(\eta_o) = \left(\frac{\tau}{2\pi D}\right)^{1/2}\exp\left(-\frac{\tau\eta_o^2}{2D}\right),$$

that yields

$$P(q,t|q_o,t_o) = \int d\eta d\eta_o P_{st}(\eta_o)P(q,\eta,t|q_o,\eta_o,t_o)$$

$$= \int_{q_o}^{q}\mathcal{D}[q(s)]\mathcal{D}[\hat{q}(s)]\exp[-\mathcal{A}[q(s),\hat{q}(s)]], \quad (7.18)$$

where

$$\mathcal{A}[q(s),\hat{q}(s)] = -i\int ds\hat{q}(s)[\dot{q}(s) - f(q(s))] + \frac{1}{2}\int ds\int ds'\hat{q}(s)\bar{C}(s,s')\hat{q}(s), \tag{7.19}$$

and

$$\bar{C}(s,s') = g(q(s))C(s,s')g(q(s')). \tag{7.20}$$

The same result was obtained by Phytian (Phytian (1977, 1980)) using a similar approach than the one we indicated in the first route. The above results are valid for any Gaussian noise. The final integration over $\hat{q}(s)$ leads to an expression similar to (7.12), but with the inverse of $\bar{C}(s,s')$ instead of the inverse of the noise correlation function.

It is also possible to include preparation effects, or to discuss the case of having a colored plus a white noise. We will not pursue this aspects here, and refer the interested reader to the original references (Colet *et al.* (1989); Wio *et al.* (1989)).

One can wonder about how to get an effective Markovian approximation from the Lagrangian-like function (7.11). As was discussed in Castro *et al.* (1995a), we can just neglect the non Markovian contributions (terms with $\ddot{q}$) as well as those non Fokker–Planck ones (that is terms with powers of $\dot{q}$ higher than 2). The resulting effective Lagrangian becomes

$$\mathcal{L}_{eff} \approx \frac{1}{4Dg(q)^2}\left\{\dot{q}\left[1 - \tau f'(q) + \frac{\tau g'(q)}{g(q)}f(q)\right] - f(q)\right\}^2, \tag{7.21}$$

from which we can obtain the form of the effective Fokker–Planck equation, that agrees with the UCNA (Jung and Hänggi (1987, 1989)). The analysis to obtain the Fokker–Planck equation will be discussed in the next chapter in relation with non-Markovian and non-Gaussian process. The physical meaning of neglecting the above indicated contributions is still missing.

7.3 The Stationary Distribution

Here we address the question of calculating the stationary distribution $P_{st}(q)$ for the process $q(t)$, starting from the Lagrangian-like function in (7.11). For simplicity we restrict ourselves to the case $g(q) = 0$. We use the definition (Graham (1978, 1987))

$$\Phi[q] = -\lim_{D\to 0} D\ln P_{st}(q). \tag{7.22}$$

In the white noise case, it is well known that $\Phi[q]$ satisfies a Hamilton–Jacobi-like equation associated to the system's Hamiltonian-like function (Graham (1978, 1987)). In our case, $\mathcal{L}$ properly speaking is not a true Lagrangian function. However, the variational problem solved by the Hamilton–Jacobi equation can be generalized to Lagrangian-like functions which depends on time derivatives of $q(t)$ of order higher than 1 (Buchdahl (1988)). The basic idea of this approach is to introduce generalized conjugate momenta associated with q and its time derivatives. In our case one introduces momenta Π^o conjugate to q, and Π^1 conjugate to $\dot{q}$ as

$$\Pi_o = \frac{\partial \mathcal{L}}{\partial \dot{q}} - \frac{d}{dt}\frac{\partial \mathcal{L}}{\partial q}, \tag{7.23}$$

$$\Pi_1 = \frac{\partial \mathcal{L}}{\partial \ddot{q}}, \tag{7.24}$$

and the generalized Hamiltonian becomes

$$\begin{aligned}\mathcal{H}(q, \dot{q}, \Pi_o, \Pi_1) &= \Pi_o \dot{q} + \Pi_1 \ddot{q} - \mathcal{L} \\ &= \Pi_o \dot{q} + \frac{1}{\tau^2}(\Pi_1)^2 - \frac{1}{\tau}\left([1 - \tau f']\dot{q} - f(q)\right)\Pi_1. \end{aligned} \tag{7.25}$$

The Hamilton–Jacobi-like equation is now obtained in the usual way. That is, replacing Π_o by $\partial\Phi/\partial q$ and Π_1 by $\partial\Phi/\partial\dot{q}$ into the Hamiltonian, and equating it to zero. Explicitly we have

$$\dot{q}\frac{\partial \Phi}{\partial q} + \frac{1}{\tau^2}\left(\frac{\partial \Phi}{\partial \dot{q}}\right)^2 - \frac{1}{\tau}\Big([1 - \tau f']\dot{q} - f(q)\Big)\frac{\partial \Phi}{\partial \dot{q}} = 0, \tag{7.26}$$

which is an equation for $\Phi[q, \dot{q}]$. We are interested in $\Phi[q, \dot{q}]$ at the value $\dot{q}_o$ that makes it minimum

$$\left.\frac{\partial \Phi}{\partial \dot{q}}\right|_{\dot{q}=\dot{q}_o} = 0 \tag{7.27}$$

An important point here is that, for the linear case (i.e. $f(q) \approx q$), the point $\dot{q} = 0$ is precisely the value $\dot{q}_o$ for which $\frac{\partial \Phi}{\partial \dot{q}} = 0$. Hence, it seems natural to propose

$$\Phi[q, \dot{q}] = \Phi_o[q] + \sum_{n=2}^{\infty} \dot{q}^n \Phi_n[q]. \tag{7.28}$$

Replacing (7.28) into (7.26) we obtain a series of equations

$$\frac{d\Phi}{dq} = -\frac{2}{\tau} f(q)\Phi_2, \tag{7.29}$$

$$0 = (\Phi_2)^2 - \frac{2}{\tau}(1 - \tau f')\Phi_2 + \frac{3}{\tau} f(q)\Phi_3 \tag{7.30}$$

$$\ldots \tag{7.31}$$

that defines the solution around the value $\dot{q} = 0$. A natural approximation seems to be the quadratic one, that is $\Phi[q, \dot{q}] \simeq \Phi_o + \dot{q}^2 \Phi_2$. In such a case we obtain

$$\Phi_2 = \frac{2}{\tau}(1 - \tau f'). \tag{7.32}$$

The consistency of the quadratic approximation requires that $\Phi_2[q] > 0$, in order that $\dot{q} = 0$ be a minimum. This is fulfilled in the case of a single attractor, with $f' < 0$. For a more detailed discussion we again refer the reader to the original references (Colet *et al.* (1989); Wio *et al.* (1989))

7.4 The Interpolating Scheme

Here we present an extension of the UCNA through its interpretation as an interpolation procedure between the white noise limit and the infinite correlation time limit. Such an interpretation can be more easily seen using a path integral description of the problem. The advantages of this procedure lie in the possibility of designing the interpolating function that best fits a particular set of experimental data, and in this way accurately predicting other relevant functions.

As indicated before, the problem of colored noise can be introduced considering a relevant macrovariable $q(t)$ which satisfies a stochastic differential equation of the form (7.1), with $g(q) = 1$.

As we have seen, it is possible to rewrite the problem as a pair of equations with a white noise term acting on one of them, that is to consider

$$\dot{q}(t) = f(q(t)) + \epsilon(t), \tag{7.33}$$

and

$$\dot{\epsilon}(t) = -\epsilon(t)\tau^{-1} + D^{-1/2}\tau^{-1}\xi(t), \tag{7.34}$$

where $\xi(t)$ is the white noise source with zero mean and δ correlated.

The UCNA results are obtained differentiating (7.33) and replacing (7.34) in it, setting $\ddot{q} = 0$ through an adiabatic elimination scheme, and making a scaling of the time variable according to $\underline{t} = t\tau^{-1/2}$. The result is a multiplicative Markovian process described by

$$\dot{q}(t) = f(q)\gamma(q, \tau)^{-1} + D^{1/2}\tau^{-1/4}\gamma(q, \tau)^{-1}\Gamma(\underline{t}), \tag{7.35}$$

where $\gamma(q, \tau) = [1 - \tau f'(q)]\tau^{-1/2}$, and $\Gamma(\underline{t})$ is an effective white noise (the prime denotes differentiation with respect to q). The Fokker–Planck equation (FPE) associated with the Langevin-like equation (7.35) has the form

$$\frac{\partial}{\partial \underline{t}} P(q,\tau,\underline{t}) = -\frac{\partial}{\partial q}\Big([f(q)\gamma(q,\tau)^{-1} - D\tau^{-1/2}\gamma'(q,\tau)\gamma(q,\tau)^{-3}]P(q,\tau,\underline{t})\Big)$$

$$+D\frac{\partial^2}{\partial q^2}\Big([\gamma(q,\tau)^{-2}\tau^{-1/2}]P(q,\tau,\underline{t})\Big). \tag{7.36}$$

The results indicated above configure the UCNA (Jung and Hänggi (1987, 1989)).

We now return to (7.33) and (7.34). It is possible to find the exact behavior for (7.33) in two limits: $\tau \to 0$ (white noise), and $\tau \to \infty$. The results for each case are

- $\tau \to 0$: the equation reduces to

$$\dot{q}(t) = f(q(t)) + D^{1/2}\xi(t), \tag{7.37}$$

with the associated FPE

$$\frac{\partial}{\partial t} P_0(q,t) = -\frac{\partial}{\partial q}\Big(f(q(t))P_0(q,t)\Big) + D\frac{\partial^2}{\partial q^2}\Big(P_0(q,t)\Big). \tag{7.38}$$

- $\tau \to \infty$: following a procedure similar to the one used for the UCNA we get

$$\dot{q}(t) = -f(q(t))\{\tau f'(q(t))\}^{-1} - D^{1/2}\{\tau f'(q(t))\}^{-1}\xi(t), \tag{7.39}$$

and the associated FPE

$$\frac{\partial}{\partial t} P_\infty(q,t) = -\frac{\partial}{\partial q}\Big(\Big\{-f(q)[\tau f'(q)]^{-1}$$

$$-D[\tau^2 f'^3(q)]^{-1} f''(q)\Big\}\, P_\infty(q,t)\Big)$$

$$+D\frac{\partial^2}{\partial q^2}\Big([\tau f'(q)]^{-2} P_\infty(q,t)\Big). \tag{7.40}$$

From a *path–integral* point of view (Langouche *et al.* (1982)), the *Lagrangian* associated with each of the Langevin ((7.37) or (7.39)) or FPE ((7.38) or (7.40)) equations are

- $\tau \to 0$

$$\mathcal{L}_0(q,\dot{q}) = \frac{1}{4D}[\dot{q} - f(q)]^2 + \frac{1}{2}f'(q). \tag{7.41}$$

- $\tau \to \infty$

$$\mathcal{L}_\infty(q,\dot{q}) = \frac{1}{4D}[\tau f'(q)\dot{q} + f(q)]^2 - \frac{1}{2\tau}. \tag{7.42}$$

Similarly, the Lagrangian associated with the UCNA ((7.35), (7.36)) has the form

$$\mathcal{L}_{UCNA}(q,\dot{q}) = \frac{1}{4D}[(1-\tau f'(q))\dot{q} - f(q)]^2 + \frac{1}{2}[1-\tau f'(q)]^{-1}f'(q). \tag{7.43}$$

Considering (7.41) and (7.42) simultaneously, it is clear that, if we have a function $\theta[\tau f'(q)]$ fulfilling the limit conditions

$$\lim_{\tau\to 0}\theta[\tau f'(q)] = 1, \tag{7.44}$$

and

$$\lim_{\tau\to\infty}\theta[\tau f'(q)] = -[\tau f'(q)]^{-1}, \tag{7.45}$$

we could define an *Interpolating Lagrangian* as

$$\mathcal{L}_I(q,\dot{q}) = \frac{1}{4D}\left\{\frac{\dot{q}}{\theta[\tau f'(q)]} - f(q)\right\}^2 + \frac{1}{2}\theta[\tau f'(q)]f'(q). \tag{7.46}$$

This Lagrangian $\mathcal{L}_I$ in the limits $\tau \to 0$ or $\tau \to \infty$, coincides with $\mathcal{L}_0(q,\dot{q})$ or $\mathcal{L}_\infty(q,\dot{q})$, respectively. The corresponding FPE is

$$\frac{\partial}{\partial t}[P_I(q,t)] = -\frac{\partial}{\partial q}\Big(\{f(q)\theta[\tau f'(q)] + D\theta[\tau f'(q)]\theta'[\tau f'(q)]\}\, P_I(q,t)\Big)$$
$$+D\frac{\partial^2}{\partial q^2}\Big(\theta^2[\tau f'(q)]P_I(q,t)\Big). \tag{7.47}$$

In this scheme, the UCNA can be interpreted as an interpolation in the sense indicated by (7.46) and (7.47), where

$$\theta[\tau f'(q)] = [1-\tau f'(q)]^{-1}. \tag{7.48}$$

This alternative point of view opens new possibilities for finding better Markovian approximations to the colored noise problem. This can be achieved through the definition of an interpolating function different from that of the UCNA, and better suited to describe the dynamics of the system.

In the next subsection we present an example of a particular family of interpolating functions, that, in some limit, reduces to the UCNA.

7.4.1 *Stationary Distributions*

In order to analyze the possibilities of the interpolation scheme, we will consider the problem of diffusion in a bistable potential driven by colored noise. We choose the very well known symmetric potential given by the expression

$$V(q) = -\frac{1}{2}aq^2 + \frac{1}{4}bq^4 \tag{7.49}$$

We introduce dimensionless variables $q \to \left(\frac{bq^2}{a}\right)^{1/2}$, $\epsilon \to \left(\frac{b\epsilon^2}{a^3}\right)^{1/2}$, $t \to at$, $D \to \frac{bD}{a^2}$; and consider $a = 1$, $b = 1$.

Remembering that the deterministic force is

$$f(q) = -\frac{dV(q)}{dq}, \tag{7.50}$$

(8.14) adopts the form

$$\dot{q} = q - q^3 + \epsilon(t). \tag{7.51}$$

We consider the family of interpolating functions given by

$$\theta[\tau f'(q)] = \frac{1 - c[\tau f'(q)]^{n-1}}{1 + c[\tau f'(q)]^n}, \tag{7.52}$$

which fulfills the requirements imposed on the limits $\tau \to 0$ and $\tau \to \infty$. In the case that $c = -1$, and $n = 2$, it reduces to the UCNA function (7.48). In the example, discussed in (Castro *et al.* (1995a)) it was only considered $n = 2$ and different values of c ranging from -1 to 1.

As the numerical procedure introduced in (Salem and Wio (1986); Abramson *et al.* (1991a)) was exploited, it was convenient to simplify the form of the potential when calculating the value of the interpolating function. It was considered a second order approximation of the original potential (7.49).

$$U(q) = \begin{cases} \frac{1}{4}\left[\frac{\sqrt{3}}{\sqrt{3}-1}(q+1)^2 - 1\right] & q \leq -\frac{1}{\sqrt{3}} \\ -\frac{1}{4}\left[\sqrt{3}q^2\right] & -\frac{1}{\sqrt{3}} < q < \frac{1}{\sqrt{3}} \\ \frac{1}{4}\left[\frac{\sqrt{3}}{\sqrt{3}-1}(q-1)^2 - 1\right] & q \geq \frac{1}{\sqrt{3}} \end{cases} \tag{7.53}$$

With this approximation, the interpolating function becomes

$$\theta(\tau) = \begin{cases} \theta_1(\tau) \mid q \mid \geq \frac{1}{\sqrt{3}} \\ \\ \theta_2(\tau) \mid q \mid < \frac{1}{\sqrt{3}} \end{cases} \tag{7.54}$$

with

$$\theta_1(\tau) = \frac{1 - c\left[\frac{\sqrt{3}\tau}{2(1-\sqrt{3})}\right]}{1 + c\left[\frac{\sqrt{3}\tau}{2(1-\sqrt{3})}\right]^2}, \tag{7.55}$$

and

$$\theta_2(\tau) = \frac{1 - c\left[\frac{\sqrt{3}\tau}{2}\right]}{1 + c\left[\frac{\sqrt{3}\tau}{2}\right]^2}. \tag{7.56}$$

We show the results obtained for the stationary probability distribution (SPD); the mean first passage time has been discussed in Castro *et al.* (1995a).

No exact expression of the stationary distribution is so far known in the colored noise problem. However, with the interpolation scheme, an approximation can be obtained. In fact, the SPD can be easily found for the FPE (7.47)

$$P_{st}(q,\tau) = \frac{N}{\theta[\tau f'(q)]} \exp\left[\frac{1}{D}\int^q \frac{f(\zeta)}{\theta[\tau f'(\zeta)]} d\zeta\right]. \tag{7.57}$$

With the interpolating function (7.54), and integrating we obtain

$$P_{st}(q,\tau) = \begin{cases} \frac{N}{2D\theta_2^2(\tau)} \exp\left[-\frac{V(q)}{D\theta_1(\tau)} + \frac{5}{36D}\left(\frac{1}{\theta_2} - \frac{1}{\theta_1}\right)\right] \mid q \mid \geq \frac{1}{\sqrt{3}} \\ \\ \frac{N}{2D\theta_2^2(\tau)} \exp\left[-\frac{V(q)}{D\theta_2(\tau)}\right] \qquad \mid q \mid < \frac{1}{\sqrt{3}} \end{cases} \tag{7.58}$$

where N is a normalization constant.

Evaluation of the SPD for different noise intensities (D) and correlation times (τ), and for different values of c were done, and compared with exact numerical results of Hänggi *et al.* (1989) and with those of UCNA (see for instance Fig. 7.1). In each case, it was found a nice agreement between the theoretical results and the numerical ones. More details can be found in Castro *et al.* (1995a). As in the case of the UCNA, the interpolation

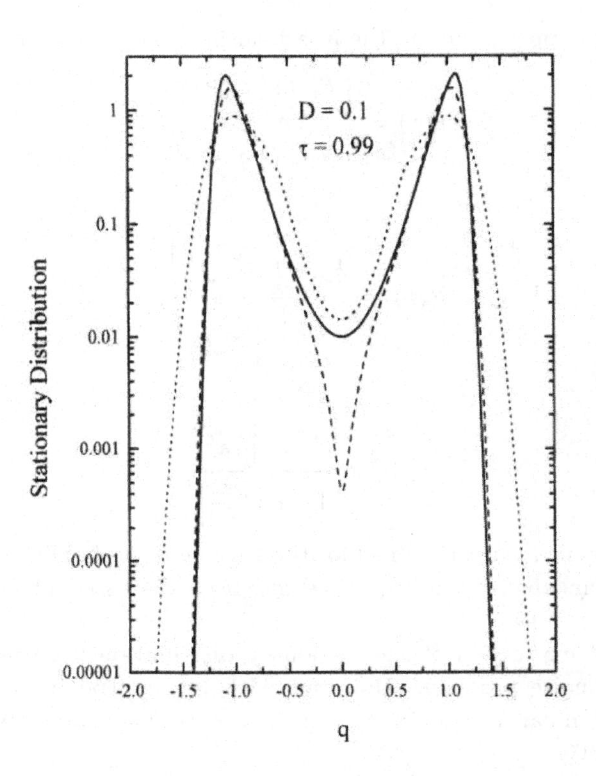

Fig. 7.1 Stationary distribution for $D = 0.1$ and $\tau = 0.99$. Exact results are indicated by a solid line, UCNA prediction with a dashed line, while the interpolation (in this case with $c = 0.44$) with a dotted line.

scheme fails to describe the stationary probability distribution in those spatial regions or for those particular values of c or τ which give a negative value of $\theta[\tau f'(q)]$, but this can be controlled with a suitable choice of c.

A nice application of the present approach to a genetic model undergoing a noise-induced transition with reentrance effects can be found in Castro *et al.* (1995b). Another application is related to a noise-induced-*phase*-transition (Mangioni *et al.* (1997, 2000)).

Chapter 8

Non-Markov Processes: Non-Gaussian Case

8.1 Introduction

As was discussed in the previous chapter, the consideration of noise sources with finite correlation time (i. e. colored noise) has become a subject of current study in the context of realistic models of physical systems.

Although, for the sake of mathematical simplicity, a majority of studies only treats the white-noise case it is expected that, because of their nature, fluctuations coupled *multiplicatively* to the system will show some degree of time-correlation or *color*, and hence give rise to new effects. A few examples are: reentrant behavior as a consequence of color in a noise-induced transition (Castro *et al.* (1995b)), an ordering nonequilibrium phase transition induced in a Ginzburg–Landau model by varying the correlation time of the additive noise, and a new reentrant phenomenon found in a simple model of noise induced phase transitions when multiplicative colored noise is considered (Mangioni *et al.* (1997)).

In a different context, most studies of the phenomenon of *stochastic resonance* (SR) have been done assuming white noise sources, with a few exceptions that studied the effect of colored noises (Gammaitoni *et al.* (1998)). In all cases the noises are assumed to be Gaussian (Gardiner (2009)). However, some experimental results in sensory systems, particularly for one kind of crayfish (Moss (1992)) as well as recent results for rat skin (Nozaki, *et al.* (1999)), offer strong indications that the noise source in these systems could be non Gaussian (Bezrukov and Vodyanoy (1995); Manwani and Koch (1999); Manwani (2000)). This point of view is supported by the results obtained in a contribution (Fuentes, *et al.* (2001a,b)) where the study of a particular class of Langevin equations having non-Gaussian stationary distribution functions (Borland (1998a)) was used.

Here we present a treatment of the problem of a non-Gaussian colored noise. The work in Borland (1998a) is based on the generalized thermostatistics proposed by Tsallis (Tsallis (1988); Curado and Tsallis (1991a,b, 1992)) which has been successfully applied to a wide variety of physical systems (Gell-Mann and C. Tsallis (2003)). We exploit a scheme similar to the one discussed in the previous chapter. Such a procedure allows us to obtain an *effective Markovian* approximation to the original non-Markovian and non-Gaussian problem. The aim is to obtain analytical expressions for some relevant quantities (particularly the mean first passage time) that could be exploited to predict qualitatively the behavior, due to the presence of a non-Gaussian colored noise, in many relevant systems.

We consider the following problem

$$\dot{x} = f(x) + g(x)\eta(t) \tag{8.1}$$

$$\dot{\eta} = -\frac{1}{\tau}\frac{d}{d\eta}V_q(\eta) + \frac{1}{\tau}\xi(t) \tag{8.2}$$

where $\xi(t)$ is a Gaussian white noise of zero mean and correlation $\langle\xi(t)\xi(t')\rangle = 2\,D\delta(t-t')$, $V_q(\eta)$ is given by (Borland (1998a,b))

$$V_q(\eta) = \frac{1}{\beta(q-1)}\ln\left[1+\beta(q-1)\frac{\eta^2}{2}\right], \tag{8.3}$$

where $\beta = \frac{\tau}{D}$. The function $f(x)$ is derived from a double well potential $U(x)$, $f(x) = -dU(x)/dx = -U'(x)$. This problem corresponds to the case of diffusion in a potential $U(x)$, induced by η, a colored non-Gaussian noise. Clearly, when $q \to 1$ we recover the limit of η being a Gaussian colored noise.

8.2 Non-Gaussian Process η

Here, and in order to determine the properties of the process η and the range of validity of the present study, we briefly analyze the stochastic process characterized by the Langevin equation given in (8.2), that is

$$\dot{\eta} = -\frac{1}{\tau}\frac{d}{d\eta}V_q(\eta) + \frac{1}{\tau}\xi(t). \tag{8.4}$$

This has the following associated Fokker–Planck equation (FPE) for the time-dependent probability density function $P_q(\eta,t)$

$$\partial_t P_q(\eta,t) = \frac{1}{\tau}\partial_\eta\left(P_q\frac{dV_q}{d\eta}\right) + \frac{D}{2\,\tau^2}\partial_\eta^2 P_q. \tag{8.5}$$

It turns out that the stationary distribution $P_q^{st}(\eta)$ is only well defined for $q \in (-\infty, 3)$, whereas for $q \geq 3$, $P_q^{st}(\eta)$ cannot be normalized and can not be accepted as a true probability function. The final expression for $P_q^{st}(\eta)$ depends on q

- For $q \in (1,3)$, we obtain a Tsallis q-exponential type form:

$$P_q^{st}(\eta) = \frac{1}{Z_q}\left[1 + \beta(q-1)\frac{\eta^2}{2}\right]^{\frac{-1}{q-1}}, \forall \eta \in (-\infty, \infty), \quad (8.6)$$

  with the normalization

$$Z_q = \int_{-\infty}^{\infty} d\eta \left[1 + \beta(q-1)\frac{\eta^2}{2}\right]^{\frac{-1}{q-1}} = \sqrt{\frac{\pi}{\beta(q-1)}}\frac{\Gamma\left(\frac{1}{q-1} - \frac{1}{2}\right)}{\Gamma\left(\frac{1}{q-1}\right)}. \quad (8.7)$$

  (Γ indicates the Gamma function). The asymptotic behavior is $P_q^{st}(\eta) \sim \eta^{\frac{-2}{q-1}}$ for $|\eta| \to \infty$. As anticipated, for $q \geq 3$ the normalization factor, Z_q, diverges.

- For $q = 1$ we recover the Gaussian distribution

$$P_1^{st}(\eta) = \frac{1}{Z_1}\exp\left(-\beta\frac{\eta^2}{2}\right), \quad (8.8)$$

  with $Z_1 = \sqrt{\pi/\beta}$. In this case of $q = 1$ the process η is nothing but an Ornstein-Uhlenbeck noise.

- Finally, for $q \in (-\infty, 1)$ we obtain a cut-off distribution, namely

$$P_q^{st}(\eta) = \begin{cases} \frac{1}{Z_q}\left[1 - \left(\frac{\eta}{w}\right)^2\right]^{\frac{1}{1-q}} & |\eta| < w \\ 0 & |\eta| > w, \end{cases}$$

  with the cut-off value given by $w = [(1-q)\frac{\beta}{2}]^{-1/2}$, and the normalizing factor being

$$Z_q = \int_{-w}^{w} d\eta \left[1 - \left(\frac{\eta}{w}\right)^2\right]^{\frac{1}{1-q}} = \sqrt{\frac{\pi}{\beta(1-q)}}\frac{\Gamma\left(\frac{1}{1-q} + 1\right)}{\Gamma\left(\frac{1}{1-q} + \frac{3}{2}\right)} \quad (8.9)$$

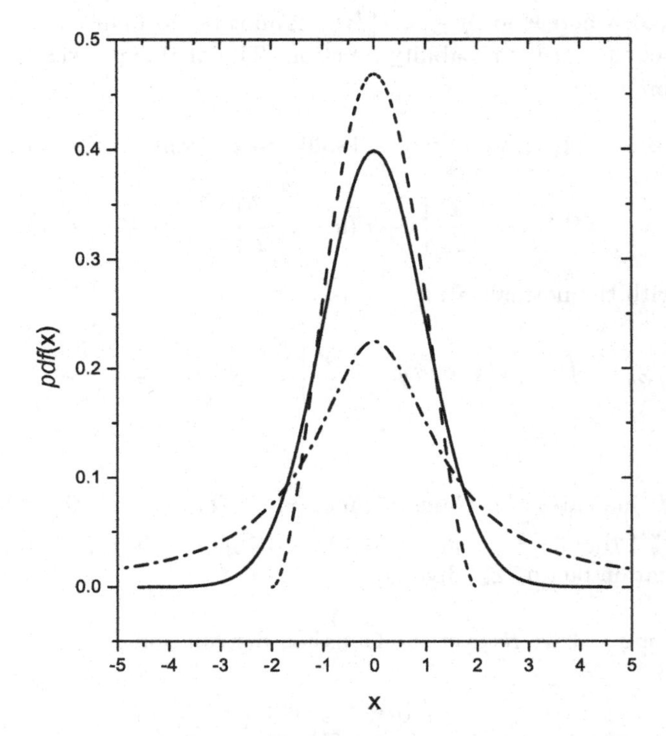

Fig. 8.1 Probability distribution $P_q^{st}(\eta)$, for different values of q: continuous line $q = 1$, dashed line $q < 1$, while the dotted-dashed is for $q > 1$.

The distribution $P_q^{st}(\eta)$ is an even function and therefore the mean value is $\langle \eta \rangle = 0$. It can be easily verified that the second moment $\langle \eta^2 \rangle$ is finite for $q < \frac{5}{3}$ and diverges for $q \in [\frac{5}{3}, 3)$. More precisely we have

$$\langle \eta^2 \rangle = \begin{cases} \frac{2}{\beta(5-3q)} & q \in (-\infty, \frac{5}{3}) \\ \\ \infty & q \in [\frac{5}{3}, 3) \end{cases}$$

In order to characterize even further the stochastic process η, we could consider its normalized time correlation function $C(t) = \langle \eta(t + t')\eta(t') \rangle / \langle \eta^2 \rangle$, in the stationary regime $t' \to \infty$. It is not possible to obtain exact analytical expressions for $C(t)$. However, given the form of (8.4) it is possible to scale out the parameters τ and D. Defining $z = \sqrt{\frac{\tau}{D}}\eta$ and $s = t/\tau$, we arrive at an equation independent of τ and D:

$$\frac{dz}{ds} = \frac{-z}{1+(q-1)\frac{z^2}{2}} + \hat{\xi}(s), \tag{8.10}$$

with $\langle \hat{\xi}(s)\hat{\xi}(s')\rangle = 2\delta(s-s')$. This leads to $C(t) = C_q(t/\tau)$ where $C_q(s) = \langle z(s+s')z(s')\rangle / \langle z^2 \rangle$ is a universal function depending only on the parameter q. In the case $q = 1$, process z is an Ornstein–Uhlenbeck noise and the correlation function is easily obtained as $C_1(s) = \exp(-s)$. It was numerically observed that this exponential decay of the correlations is still valid for $q < 1$ where we can write $C_q(s) = \exp(-s/s_q)$. This exponential behavior fails for $q > 1$ where, on the other side, $C_q(s)$ can be approximated by a Tsallis q-like exponential (Tsallis (1988); Curado and Tsallis (1991a,b, 1992)) $C_q(s) = [1+(q-1)\frac{s}{s_q}]^{\frac{1}{1-q}}$. For instance, the characteristic correlation time s_q defined as

$$s_q = \int_0^\infty ds C_q(s), \tag{8.11}$$

is such that it diverges for $q \to 5/3$, as $s_q = 2/(5-3q)$. Notice that in the limit $q \to 1$ it gives the exact result $s_1 = 1$ (corresponding to the Ornstein–Uhlenbeck process). Although it was not possible to derive this result analytically, a very simple calculation is able to predict the divergence of s_q for $q = q_c = 5/3$. We have

$$\frac{dC_q(s)}{ds} = -\left\langle \frac{z(s)z(0)}{1+(q-1)\frac{z(s)^2}{2}} \right\rangle \approx -\frac{1}{s_q} C_q(s), \tag{8.12}$$

where we have approximated

$$s_q \approx \frac{1}{1+(q-1)\frac{\langle z^2 \rangle}{2}} = \frac{2(2-q)}{5-3q}, \tag{8.13}$$

which indeed diverges as $s_q \sim (5-3q)^{-1}$ although with a different prefactor from the one observed numerically.

8.3 Effective Markov Approximation

As indicated before, applying the path-integral formalism to the Langevin equations given in (8.1),(8.2), and making an adiabatic like elimination procedure it is possible to arrive to an *effective Markovian approximation*. Here we sketch such a procedure.

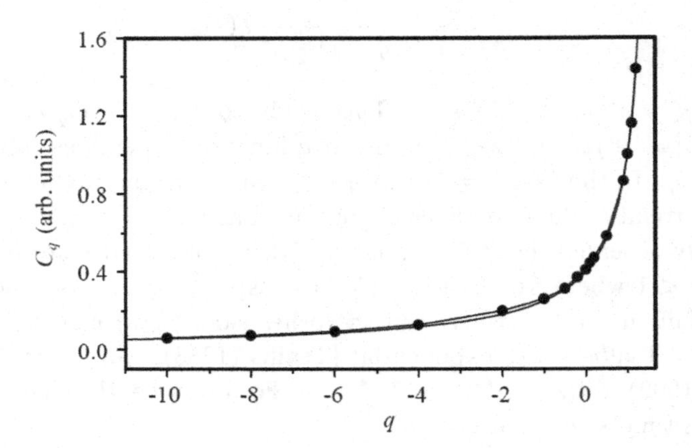

Fig. 8.2 Correlation time s_q of $C_q(s)$ (the correlation function of the process η), as a function of q. Simulations are indicated by dots, while the theory with the continuous line.

The FPE associated to (8.1) and (8.2) is

$$\begin{aligned}\frac{\partial}{\partial t}P_q(x,\eta,\tau,t) = &-\frac{\partial}{\partial x}\Big([f(x)+g(x)\eta]P_q(x,\eta,\tau,t)\Big)\\ &+\frac{\partial}{\partial \eta}\Big(\tau^{-1}[\frac{d}{d\eta}V_q(\eta)]P_q(x,\eta,\tau,t)\Big)\\ &+\frac{D}{\tau^2}\frac{\partial^2}{\partial \eta^2}P_q(x,\eta,\tau,t).\end{aligned} \tag{8.14}$$

The phase space path integral representation for the transition probability, corresponding to the Langevin equations given in (8.1),(8.2) or to the associated FPE in (8.14) is

$$\begin{aligned}&P_q(x_b,\eta_b,t_b \mid x_a,\eta_a,t_a;\tau) =\\ &= \int_{x(t_a)=x_a,\eta=\eta_b}^{x(t_b)=x_b,\eta=\eta_a} \mathcal{D}\,[x(t)]\;\mathcal{D}\,[\eta(t)]\;\mathcal{D}\,[p_x(t)]\;\mathcal{D}\,[p_\eta(t)]\;\exp\,[\mathcal{S}_{q,1}],\end{aligned} \tag{8.15}$$

where $p_x(t)$ and $p_\eta(t)$ are the canonically conjugate variables to $x(t)$ and $\eta(t)$, respectively. $\mathcal{S}_{q,1}$ is the *stochastic action* given by

$$\mathcal{S}_{q,1} = \int_{t_a}^{t_b} ds \quad \Big(ip_x(s)[\dot{x}(s) - f(x(s)) - g(x(s))\eta(s)]$$
$$+ ip_\eta(s)[\dot{\eta}(s) + \tau^{-1}[\frac{d}{d\eta}V_q(\eta(s))]] + \frac{D}{\tau^2}(ip_\eta(s))^2\Big). \tag{8.16}$$

The Gaussian integration over $p_\eta(s)$ yields

$$P_q(x_b, \eta_b, t_b \mid x_a, \eta_a, t_a; \tau) = \int_{x(t_a)=x_a,\eta=\eta_b}^{x(t_b)=x_b,\eta=\eta_a} \mathcal{D}\,[x(t)]\,\mathcal{D}\,[\eta(t)]\,\mathcal{D}\,[p_x(t)]\; e^{\mathcal{S}_{q,2}}, \tag{8.17}$$

with

$$\mathcal{S}_{q,2} = \int_{t_a}^{t_b} ds\Big(ip_x(s)[\dot{x}(s) - f(x(s)) - g(x(s))\eta(s)] + \frac{\tau^2}{4D}\int_{t_a}^{t_b} ds'[\dot{\eta}(s)$$
$$+\tau^{-1}[\frac{d}{d\eta}V_q(\eta(s))]\delta(s-s')[\dot{\eta}(s') + \tau^{-1}[\frac{d}{d\eta}V_q(\eta(s'))]\Big). \tag{8.18}$$

The integration over $p_x(s)$ is also immediate, yielding

$$P_q(x_b, \eta_b, t_b \mid x_a, \eta_a, t_a; \tau) \sim \int_{x(t_a)=x_a,\eta=\eta_b}^{x(t_b)=x_b,\eta=\eta_a} \mathcal{D}\,[x(t)]\,\mathcal{D}\,[\eta(t)]$$
$$\times\hat{\delta}\,[\dot{x}(s) - f(x(s)) - g(x(s))\eta(s)]\; e^{\mathcal{S}_{q,3}}, \tag{8.19}$$

with

$$\mathcal{S}_{q,3} = \int_{t_a}^{t_b} ds\,\Big(\frac{\tau^2}{4D}\,[\dot{\eta}(s) + \tau^{-1}[\frac{d}{d\eta}V_q(\eta(s))]\Big)^2, \tag{8.20}$$

and $\hat{\delta}\,[\dot{x}(s) - f(x(s)) - g(x(s))\eta(s)]$ indicates that, at each instant of time, we have

$$\eta(s) = (\dot{x}(s) - f(x(s)))/g(x(s). \tag{8.21}$$

The last condition makes trivial the integration over $\eta(s)$. It corresponds to replacing $\eta(s)$ and $\dot{\eta}(s)$ by (8.20) and by

$$\dot{\eta}(s) = -\frac{1}{g(x)^2}\frac{dg(x)}{dx}\,\dot{x}(s)(\dot{x}(s) - f(x(s)))$$
$$+\frac{1}{g(x)}\,(\ddot{x}(s) - \frac{d}{dx}f(x(s)\,\dot{x}(s))), \tag{8.22}$$

respectively. The resulting stochastic action corresponds to a *non-Markovian* description as it involves $\ddot{x}(s)$. In order to obtain an *effective*

Markovian approximation we resort to the same kind of approximations and arguments used in relation with the colored Gaussian noise, that allows us to get a result resembling the UCNA. In simple words, such an approximation corresponds to neglecting all contributions including $\ddot{x}(s)$ and/or $\dot{x}(s)^n$ with $n > 1$. Doing this we get the approximate relation

$$\begin{aligned}\dot{\eta} + \tau^{-1}[\frac{d}{d\eta}V_q(\eta)] \approx & -\frac{1}{g(x)}\left(\frac{d}{dx}f(x)\,\dot{x} - f(x)\frac{d}{dx}\ln g(x)\dot{x}\right) \\ & + \frac{1}{\tau g(x)}\frac{\dot{x}(s) - f(x(s))}{1 + \frac{\beta(q-1)}{2}\left(\frac{f(x)}{g(x)}\right)^2} \\ & - \frac{1}{\tau g(x)}\frac{\beta(q-1)f(x)^2\dot{x}(s)}{\left(1 + \frac{\beta(q-1)}{2}\left(\frac{f(x)}{g(x)}\right)^2\right)^2}. \end{aligned} \tag{8.23}$$

As in the case of UCNA, this approximation gives reliable results for small values of τ.

The final result for the transition probability is

$$P_q(x_b, \eta_b, t_b \mid x_a, \eta_a, t_a; \tau) = \int_{x(t_a)=x_a,\eta=\eta_b}^{x(t_b)=x_b,\eta=\eta_a} \mathcal{D}\,[x(t)]\, e^{\mathcal{S}_{q,4}}, \tag{8.24}$$

with (for the simple case $g(x) = 1$)

$$\mathcal{S}_{q,4} = \frac{1}{4D}\int_{t_a}^{t_b} ds\left(\left[-\tau\frac{df(x)}{dx} + \frac{[1 - \frac{\beta(q-1)}{2}f(x)^2]}{[1 + \frac{\beta(q-1)}{2}f(x)^2]^2}\right]\dot{x} - \frac{f(x)}{[1 + \frac{\beta(q-1)}{2}f(x)^2]}\right)^2. \tag{8.25}$$

It is immediate to recover some known limits. For $\tau > 0$ and $q \to 1$ we get the known Gaussian colored noise result (Ornstein–Uhlenbeck process), while for $q \neq 1$ and $\tau \to 0$ we find the case of Gaussian white noise.

The FPE for the evolution of the probability $P(x,t)$ corresponding to the above indicated path integral representation is

$$\frac{\partial}{\partial t}P(x,t) = -\frac{\partial}{\partial x}[A(x)P(x,t)] + \frac{1}{2}\frac{\partial^2}{\partial x^2}[B(x)P(x,t)], \tag{8.26}$$

where

$$A(x) = \frac{U'}{\left(\frac{1 - \frac{\tau}{2D}(q-1)U'^2}{1 + \frac{\tau}{2D}(q-1)U'^2}\right) + \tau U''[1 + \frac{\tau}{2D}(q-1)U'^2]}, \tag{8.27}$$

and

$$B(x) = D\left(\frac{[1 + \frac{\tau}{2D}(q-1)U'^2]^2}{\tau U''[1 + \frac{\tau}{2D}(q-1)U'^2]^2 + [1 - \frac{\tau}{2D}(q-1)U'^2]}\right)^2. \tag{8.28}$$

The stationary distribution of the FPE in (8.26) results

$$P^{st}(x) = \frac{\aleph}{B} \exp\left[-\Phi(x)\right], \tag{8.29}$$

where $\aleph$ is the normalization factor, and

$$\Phi(x) = 2 \int \frac{A}{B} dy. \tag{8.30}$$

The indicated FPE and its associated stationary distribution allow us to obtain the *mean first-passage time* (MFPT). The MFPT can be obtained, in a Kramers-like approximation (Gardiner (2009)) from

$$T(x_0) = \int_a^{x_0} \frac{dy}{\Psi} \int_{-\infty}^{y} \frac{dz \Psi}{B}, \tag{8.31}$$

where

$$\Psi(x) = \exp\left(-2 \int dy \frac{A}{B}\right). \tag{8.32}$$

We will focus on polynomial-like forms for the potential and adopt

$$U(x) = \frac{x^4}{4} - \frac{x^2}{2}. \tag{8.33}$$

For this kind of potential the normalization constant $\aleph$ diverges for any value of $\tau > 0$. This can be seen from the asymptotic behavior of $\Psi(x)$, for $x \to \infty$: $\Psi(x) \to 0$, while $B^{-1} \to \infty$, resulting in a ill defined stationary probability density in (8.29). In order to find approximate relations for the MFPT, and other related quantities, we assume that (8.26) is valid only for values of x near the wells and when the dispersion of the η process is finite, that is $\langle \eta^2 \rangle < \infty$ (or $q \in (-\infty, \frac{5}{3})$). Such a cutoff is justified a posteriori, analyzing the probability distributions that can be obtained from the simulations.

In order to obtain the MFPT and related quantities, we have to integrate (8.31), replacing the potential given by (8.33) in the expressions for $A(x)$ and $B(x)$. It is worth remarking here that the relevant quantity is not the white noise intensity D but the non-Gaussian noise intensity D_{nG}. Both quantities are related through

$$D_{nG} = 2D(5 - 3q)^{-1}.$$

We will not discuss here neither the simulations nor the comparison between theoretical results and simulations, but just refer to the original references (Fuentes, *et al.* (2001a,b); Castro, *et al.* (2001); Wio (2005)).

Chapter 9

Non-Markov Processes: Nonlinear Cases

9.1 Introduction

In this chapter, and following the tendency marked in the two previous chapters, we will discuss a few cases where the noise enters in a nonlinear form. For instance, such problems were studied by Sagues (1984); Sagues *et al.* (1986); Luzcka (1988); Luzcka *et al.* (1995), analyzing different nonlinear dependencies on the noise. Also, in a recent paper (Calisto *et al.* (2006)) analyzed stochastic resonance in a linear model forced with a periodic term, and subject to a multiplicative linear combination of a colored noise and its square. These studies created the need, and motivated us, to investigate how to obtain a Markov approximations to such originally non-Markov problems.

We will consider the following kind of problems

$$\dot{x} = F[x, \eta(t)], \tag{9.1}$$

where

$$\dot{\eta} = -\frac{\eta}{\tau} + \frac{1}{\tau}\xi(t), \tag{9.2}$$

with $\xi(t)$ a Gaussian white noise of zero mean and correlation $\langle \xi(t)\xi(t')\rangle = 2D\delta(t - t')$. $F[x, \eta(t)]$ indicates a nonlinear function of x and $\eta(t)$. For this function we will consider the following forms

$$F[x, \eta(t)] = a_o(x) + a_1(x)\eta(t) + a_2(x)\eta(t)^2 + \ldots,$$

with $a_n(x)$ functions of x, and also the following exponential case

$$F[x, \eta(t)] = f(x) + g(x)\exp[\eta(t)].$$

In addition, we will also discuss the reduction of the Kramers equation (in phase space (x, p)) to a Markov description (in configuration space (x)).

9.2 Nonlinear Noise

Here we will discuss how to treat a few cases associated to the general nonlinear situations indicated above.

9.2.1 *Polynomial Noise*

We start considering the following situation

$$F[x,\eta(t)] = a_o(x) + a_1(x)\eta(t) + a_2(x)\eta(t)^2 + \ldots,$$

but, in order to simplify our treatment, we will only discuss the more simply case $F[x,\eta(t)] = a_o(x) + a_n(x)\eta(t)^n$, and $a_o(x) = a_o = cte$. Hence, the system under study will be

$$\dot{x} = a_n(x)\eta(t)^n + a_o, \tag{9.3}$$

together with (9.2).

We now sketch the procedure to obtain an *effective Markovian approximation* to the non Markovian system indicated above. The FPE associated to (9.3) and (9.2) is

$$\begin{aligned}\frac{\partial}{\partial t}P(x,\eta,t) = & -\frac{\partial}{\partial x}\Big(-[a_n(x)\eta(t)^n + a_o]P(x,\eta,t)\Big) \\ & +\frac{\partial}{\partial \eta}\Big(\frac{\eta}{\tau}P(x,\eta,t)\Big) + \frac{D}{\tau^2}\frac{\partial^2}{\partial \eta^2}P(x,\eta,t).\end{aligned} \tag{9.4}$$

The phase space path integral representation for the transition probability, corresponding to the Langevin equations given in (9.3) and (9.2) or to the associated FPE in (9.4) is

$$\begin{aligned}& P(x_b,\eta_b,t_b \mid x_a,\eta_a,t_a) = \\ & = \int_{x(t_a)=x_a,\eta=\eta_b}^{x(t_b)=x_b,\eta=\eta_a} \mathcal{D}\,[x(t)]\,\mathcal{D}\,[\eta(t)]\,\mathcal{D}\,[p_x(t)]\,\mathcal{D}\,[p_\eta(t)]\,\exp{[\mathcal{S}_1]},\end{aligned} \tag{9.5}$$

where $p_x(t)$ and $p_\eta(t)$ are the canonically conjugate variables to $x(t)$ and $\eta(t)$, respectively. $\mathcal{S}_1$ is the stochastic *action* given by

$$\begin{aligned}\mathcal{S}_1 = & \int_{t_a}^{t_b} ds\Big\{ip_x(s)[\dot{x}(s) - [a_n(x)\eta(t)^n + a_o]] \\ & +ip_\eta(s)[\dot{\eta}(s) + \frac{\eta}{\tau}] + \frac{D}{\tau^2}(ip_\eta(s))^2\Big\}.\end{aligned} \tag{9.6}$$

As was indicated in previous chapters, the Gaussian integration over $p_\eta(s)$ is immediate and yields

$$\begin{aligned}P(x_b,\eta_b,t_b \mid x_a,\eta_a,t_a) = & \int_{x(t_a)=x_a,\eta=\eta_b}^{x(t_b)=x_b,\eta=\eta_a} \\ & \mathcal{D}\,[x(t)]\,\mathcal{D}\,[\eta(t)]\,\mathcal{D}\,[p_x(t)]\,e^{\mathcal{S}_2},\end{aligned} \tag{9.7}$$

with

$$\mathcal{S}_2 = \int_{t_a}^{t_b} ds \Big\{ ip_x(s)[\dot{x}(s) - [a_n(x)\eta(t)^n + a_o]] + \frac{\tau^2}{4D}[\dot{\eta}(s) + \frac{\eta}{\tau}]^2 \Big\}. \tag{9.8}$$

The integration over $p_x(s)$ is also immediate, yielding a δ-like contribution

$$P(x_b, \eta_b, t_b \mid x_a, \eta_a, t_a) \sim \int_{x(t_a)=x_a,\eta=\eta_b}^{x(t_b)=x_b,\eta=\eta_a} \mathcal{D}\left[x(t)\right] \mathcal{D}\left[\eta(t)\right]$$
$$\times \hat{\delta}\left[\dot{x}(s) - (a_n(x)\eta(s)^n + a_o)\right] \, e^{\mathcal{S}_3}, \tag{9.9}$$

where the action $\mathcal{S}_3$ is

$$\mathcal{S}_3 = \int_{t_a}^{t_b} ds \left(\frac{\tau^2}{4D} \left[\dot{\eta}(s) + \frac{\eta}{\tau}\right] \right)^2, \tag{9.10}$$

and $\hat{\delta}\left[\dot{x}(s) - (a_n(x)\eta(s)^n + a_o)\right]$ indicates that, at each instant of time, we have to fulfill the condition

$$\dot{x}(s) = a_n(x)\eta(s)^n + a_o.$$

The δ condition makes trivial the integration over $\eta(s)$. It corresponds to replacing $\eta(s)$ and $\dot{\eta}(s)$ by

$$\eta(s) = \left(\frac{\dot{x}(s) - a_o}{a_n(x)} \right)^{1/n}, \tag{9.11}$$

and

$$\dot{\eta}(s) = \left(\frac{\ddot{x}(s)}{a_n(x)} - \frac{\dot{x}(s) - a_o}{a_n(x)^2} a_n'(x)\dot{x}(s) \right), \tag{9.12}$$

respectively. As occurred in the colored (Chapter 7) and non-Gaussian (Chapter 8) cases, the resulting stochastic action corresponds to a *non-Markovian* description as it involves a dependence on $\ddot{x}(s)$

$$P(x_b, \eta_b, t_b \mid x_a, \eta_a, t_a) = \int_{x(t_a)=x_a,\eta=\eta_b}^{x(t_b)=x_b,\eta=\eta_a} \mathcal{D}\left[x(t)\right] e^{\mathcal{S}_4}, \tag{9.13}$$

with

$$\mathcal{S}_4 = \int_{t_a}^{t_b} ds \, \frac{1}{4D} \Big\{ \tau\dot{\eta}[x(s), \dot{x}(s), \ddot{x}(s)] + \eta[x(s), \dot{x}(s), \ddot{x}(s)] \Big\}^2. \tag{9.14}$$

In order to obtain an *effective* Markovian approximation we should resort again to the same kind of approximations and arguments used in relation with the previous colored Gaussian and non-Gaussian noise cases. In simple words, such an approximation corresponds to neglecting all contributions including $\ddot{x}(s)$ and/or $\dot{x}(s)^n$ with $n > 1$.

9.2.2 *Exponential Noise*

Here, we consider now the second situation, that is the exponential nonlinearity

$$F[x,\eta(t)] = f(x) + g(x)\exp[\eta(t)].$$

Hence the system to be studied results to be

$$\dot{x} = f(x) + g(x)\exp[\eta(t)], \tag{9.15}$$

together with (9.2).

We again sketch the procedure to obtain an *effective Markovian approximation* to the non-Markovian system indicated above. The FPE associated to (9.15) and (9.2) is

$$\begin{aligned}\frac{\partial}{\partial t}P(x,\eta,t) = & -\frac{\partial}{\partial x}\Big(-[f(x) + g(x)\exp[\eta(t)]]P(x,\eta,t)\Big)\\ & + \frac{\partial}{\partial \eta}\Big(\frac{\eta}{\tau}P(x,\eta,t)\Big) + \frac{D}{\tau^2}\frac{\partial^2}{\partial \eta^2}P(x,\eta,t). \end{aligned} \tag{9.16}$$

As before, we look for the phase space path integral representation for the transition probability, corresponding to the Langevin equations given in (9.15) and (9.2) or to the associated FPE in (9.16). It is given by

$$\begin{aligned} & P(x_b,\eta_b,t_b \mid x_a,\eta_a,t_a) = \\ & = \int_{x(t_a)=x_a,\eta=\eta_b}^{x(t_b)=x_b,\eta=\eta_a} \mathcal{D}\,[x(t)]\;\mathcal{D}\,[\eta(t)]\;\mathcal{D}\,[p_x(t)]\;\mathcal{D}\,[p_\eta(t)]\;\exp[\mathcal{S}_1], \end{aligned} \tag{9.17}$$

where, as before, $p_x(t)$ and $p_\eta(t)$ are the canonically conjugate variables to $x(t)$ and $\eta(t)$, respectively. Also, $\mathcal{S}_1$ is the stochastic *action* given by

$$\begin{aligned}\mathcal{S}_1 = \int_{t_a}^{t_b} ds \Big\{ & ip_x(s)[\dot{x}(s) - [f(x) + g(x)\exp[\eta(t)]]]\\ & + ip_\eta(s)[\dot{\eta}(s) + \frac{\eta}{\tau}] + \frac{D}{\tau^2}(ip_\eta(s))^2\Big\}. \end{aligned} \tag{9.18}$$

Again, the Gaussian integration over $p_\eta(s)$ is immediate, yielding

$$P(x_b,\eta_b,t_b \mid x_a,\eta_a,t_a) = \int_{x(t_a)=x_a,\eta=\eta_b}^{x(t_b)=x_b,\eta=\eta_a} \mathcal{D}\,[x(t)]\;\mathcal{D}\,[\eta(t)]\;\mathcal{D}\,[p_x(t)]\;e^{\mathcal{S}_2}, \tag{9.19}$$

with

$$\mathcal{S}_2 = \int_{t_a}^{t_b} ds \Big\{ i p_x(s)[\dot{x}(s) - [f(x) + g(x)\exp[\eta(t)]]] + \frac{\tau^2}{4D}[\dot{\eta}(s) + \frac{\eta}{\tau}]^2 \Big\}. \tag{9.20}$$

In the last expression for $P(x_b, \eta_b, t_b \mid x_a, \eta_a, t_a)$, with the action $\mathcal{S}_2$, the integration over $p_x(s)$ is also immediate, yielding again a δ-like contribution

$$P(x_b, \eta_b, t_b \mid x_a, \eta_a, t_a) \sim \int_{x(t_a)=x_a,\eta=\eta_b}^{x(t_b)=x_b,\eta=\eta_a} \mathcal{D}\left[x(t)\right] \mathcal{D}\left[\eta(t)\right]$$

$$\times \hat{\delta}\left[\dot{x}(s) - [f(x) + g(x)\exp[\eta(t)]]\right] e^{\mathcal{S}_3}, \tag{9.21}$$

with $\mathcal{S}_3$ having a form similar to the one indicated in (9.10) and

$$\hat{\delta}\left[\dot{x}(s) - [f(x) + g(x)\exp[\eta(t)]]\right],$$

indicating once more that, at each instant of time, we have

$$\dot{x}(s) = f(x) + g(x)\exp[\eta(t)].$$

We can use this last expression to obtain

$$\eta(t) = \ln\left[\frac{\dot{x}(s) - f(x)}{g(x)}\right], \tag{9.22}$$

that implies that only values such that

$$\left[\frac{\dot{x}(s) - f(x)}{g(x)}\right] > 0$$

will contribute as in other case the relation will have no physical meaning. From here we obtain for $\dot{\eta}(t)$

$$\dot{\eta}(t) = \frac{g(x)}{\dot{x}(s) - f(x)}\left[\frac{\ddot{x}(s) - f'(x)\dot{x}(s)}{g(x)} - \frac{\dot{x}(s) - f(x)}{g(x)^2} g'(x)\dot{x}(s)\right]. \tag{9.23}$$

Once more, exploiting the δ behavior, the integration over η can be readily done. The resulting stochastic action also corresponds to a *non-Markovian* description involving contributions with $\ddot{x}(s)$

$$P(x_b, \eta_b, t_b \mid x_a, \eta_a, t_a) = \int_{x(t_a)=x_a,\eta=\eta_b}^{x(t_b)=x_b,\eta=\eta_a} \mathcal{D}\left[x(t)\right] e^{\mathcal{S}_4}, \tag{9.24}$$

where, $\mathcal{S}_4$ is again giving by a form like in (9.10).

Again, in order to obtain an *effective* Markovian approximation we should resort to an approximation, that corresponds to neglecting all contributions including $\ddot{x}(s)$ and/or $\dot{x}(s)^n$ with $n > 1$. For the present case, the effective Markovian action will read

$$S_{Eff} \sim \int_{t_a}^{t_b} ds \, \frac{1}{4D}\Big\{\Phi(x) - (1 + \tau\frac{d\Phi(x)}{dx})\dot{x}(s)\Big\}^2, \qquad (9.25)$$

where

$$\Phi(x) = \ln\left(-\frac{f(x)}{g(x)}\right),$$

and only the range where $-\frac{f(x)}{g(x)} > 0$ will be physically accessible.

9.3 Kramers Problem

In this section I want to introduce some ideas related to having a reasonable reduced (that is, in configuration space) Markovian approximation to the Kramers equation.

The Kramers' problem corresponds to the motion of a Brownian particle in a field of force, in a phase-space description (that is the space of variables x and v), as indicated by the coupled system of stochastic differential equations (SDE)

$$\begin{aligned} \dot{x} &= v(t), \\ M\dot{v} &= -M\gamma v(t) - U'(x(t)) + \sigma\xi(t). \end{aligned} \qquad (9.26)$$

Here M is the "particle mass", the noise is a white one, such that $\langle\xi(t)\rangle = 0$ and $\langle\xi(t)\xi(t')\rangle = 2\delta(t-t')$; and $\sigma = \sqrt{2M\gamma kT}$. The associated Fokker–Planck equation (FPE) [with $P = P(x, v, t \mid x_i, v_i, t_i)$] is

$$\frac{\partial}{\partial t}P = -\frac{\partial}{\partial x}[v(t)P] + \frac{\partial}{\partial v}\left\{\left[-\gamma v(t) + \frac{U'(x(t))}{M}\right]P\right\} + \frac{\sigma^2}{M^2}\frac{\partial^2}{\partial v^2}P. \qquad (9.27)$$

As indicated in Gardiner (2009), in the overdamped limit, and within the configuration space (the space of the variable x), we obtain the so called Smoluchowski equation

$$\frac{\partial}{\partial t}\hat{P}(x, t \mid x_i, t_i) = \frac{\partial}{\partial x}\Big[\frac{U'(x(t))}{M\gamma}\hat{P}\Big] + \frac{kT}{M\gamma}\frac{\partial^2}{\partial x^2}\hat{P}, \qquad (9.28)$$

with the associated Langevin equation

$$\dot{x} = -\frac{U'(x(t))}{M\gamma} + \sqrt{\frac{kT}{M\gamma}}\xi(t). \qquad (9.29)$$

Our starting point will be the coupled system of SDEs (9.26). We seek a path-integral representation of the transition probability $P \equiv P_{M,\sigma}(x_f, v_f, t_f \mid x_i, v_i, t_i)$, which obeys the FPE (9.27). As we have seen before, the difficulties associated with the fact that this FPE has singular diffusion matrix can be circumvented by performing the functional integral in the *phase space* of the variables $x(t)$ and $v(t)$ [Langouche *et al.* (1982); Wio *et al.* (1989); Fuentes, *et al.* (2001a)]

$$P = \int_{x(t_i)=x_i, v(t_i)=v_i}^{x(t_f)=x_f, v(t_f)=v_f} \mathcal{D}\left[x(t)\right] \mathcal{D}\left[v(t)\right] \mathcal{D}\left[p_x(t)\right] \mathcal{D}\left[p_v(t)\right] e^{\mathcal{S}_1(M,\sigma)}. \tag{9.30}$$

Here, as before, $p_x(t)$ and $p_v(t)$ are the canonical momenta conjugate to $x(t)$ and $v(t)$, respectively, and $\mathcal{S}_1(M,\sigma)$ is the "stochastic action"

$$\begin{aligned} \mathcal{S}_1(M,\sigma) = \int_{t_i}^{t_f} ds \Big\{ & ip_x(s)\left[\dot{x}(s) - v(s)\right] \\ & + ip_v(s)\Big[\dot{v}(s) + \gamma v(s) + \frac{U'(x(s))}{M}\Big] + \left(\frac{\sigma}{M}\right)^2 (ip_v(s))^2 \Big\}. \end{aligned} \tag{9.31}$$

The Gaussian integration over $p_v(s)$ yields

$$P = \int_{x(t_i)=x_i, v(t_i)=v_i}^{x(t_f)=x_f, v(t_f)=v_f} \mathcal{D}\left[x(t)\right] \mathcal{D}\left[v(t)\right] \mathcal{D}\left[p_x(t)\right] e^{\mathcal{S}_2(M,\sigma)}, \tag{9.32}$$

with

$$\begin{aligned} \mathcal{S}_2(M,\sigma) = \int_{t_i}^{t_f} ds \Big\{ & -\left(\frac{M}{2\sigma}\right)^2 \left[\dot{v}(s) + \frac{U'(x(s))}{M} + \gamma v(s)\right]^2 \\ & + ip_x(s)\left[\dot{x}(s) - v(s)\right] \Big\}. \end{aligned} \tag{9.33}$$

The integration over $p_x(s)$ is immediate, and yields

$$\hat{\delta}\left[\dot{x}(s) - v(s)\right].$$

This allow us to integrate over $v(s)$, exploiting the above indicated $\hat{\delta}$. We obtain, using $v(s) = \dot{x}(s)$ and $\dot{v}(s) = \ddot{x}(s)$, that

$$P = \int_{x(t_i)=x_i}^{x(t_f)=x_f} \mathcal{D}\left[x(t)\right] e^{\mathcal{S}_3(M,\sigma)}, \tag{9.34}$$

with

$$\mathcal{S}_3(M,\sigma) = -\left(\frac{M}{2\sigma}\right)^2 \int_{t_i}^{t_f} ds \left\{ \ddot{x}(s) + \gamma \dot{x}(s) + \frac{U'(x(s))}{M} \right\}^2. \tag{9.35}$$

This stochastic action provides a *non-Markovian* description, since it involves time derivatives higher than $\dot{x}(s)$. Clearly, if we want a Markovian description we can resort to the *strong Markovian approximation*, i.e. to eliminate them (as in Fuentes, *et al.* (2001a)). That is, we eliminate $\ddot{x}(s)$ (and, if we want a FPE description, also $\dot{x}(s)^n$ with $n > 1$). The final approximate Markovian action reads

$$\mathcal{S}_4(M,\sigma) \approx -\left(\frac{M}{2\sigma}\right)^2 \int_{t_i}^{t_f} ds \left\{\gamma\dot{x}(s) + \frac{U'(x(s))}{M}\right\}^2, \qquad (9.36)$$

where all information about inertia has disappeared. As a matter of fact, the result corresponds to the Smoluchovski equation (Gardiner (2009)), that is the overdamped limit. It is clear that, if one wants to retain information about the inertia, but still within an effective Markov framework, a more elaborate approximation is required.

Chapter 10

Fractional Diffusion Process

In this chapter we will discuss a few aspects related to the path integral approach as applied to some stochastic processes exhibiting anomalous diffusion (that is being sub- or super-diffusive). We start introducing a few ideas related to the so called *fractional Brownian motion*, continue presenting a straightforward approach to the path integral representation of such a kind of processes, analyze some possible extensions, and finally discuss the case of *fractional Lévy motion*

10.1 Short Introduction to Fractional Brownian Motion

A normalized fractional Brownian motion (fBm), also called a *fractal Brownian motion*, is a continuous time Gaussian process $B_H(t)$ on the interval $[0,T]$, which starts at $\beta_H(0) = 0$, has expectation $\langle \beta_H(t) \rangle = 0$ for all t ($t \in [0,T]$), and the following covariance function

$$\langle \beta_H(t)\beta_H(s) \rangle = \frac{1}{2}\Big[|t|^{2H} + |s|^{2H} - |t-s|^{2H}\Big],$$

where H is a real number ($H \in (0,1)$), called the Hurst index, Hurst exponent, or Hurst parameter, associated with the fractional Brownian motion. The Hurst exponent describes the raggedness of the resultant motion, with a higher value leading to a smoother motion (Mandelbrot and van Ness (1968)).

The value of H determines what kind of process the fBm is

- if $H = \frac{1}{2}$ then the process is a "normal" Brownian motion or Wiener process, or normal diffusion;

- if $H > \frac{1}{2}$ then the increments of the process are positively correlated, leading to a super-diffusive motion;

- while if $H < \frac{1}{2}$ the increments of the process are negatively correlated, leading to sub-diffusion.

Some relevant properties are

- Self-similarity: fBm is the only self-similar Gaussian process. This means that, in terms of probability distributions, $\beta_H(at) \sim |a|^H \beta_H(t)$;
- Stationary increments: the increments $X(t) = \beta_H(t+1) - \beta_H(t)$ are known as fractional Gaussian noise. Such increments are stationary $\beta_H(t) - \beta_H(s) \sim \beta_H(t-s)$;
- Long-range dependence: when $H > \frac{1}{2}$ the process exhibits long-range dependence as indicated by $\sum_{n=1}^{\infty} \langle \beta_H(1)(\beta_H(n+1) - \beta_H(n)) \rangle = \infty$;
- Regularity: sample-path are almost nowhere differentiable. However, almost all trajectories are Hölder continuous for any order strictly lower than H. That is, for each trajectory exists a constant c such that $|\beta_H(t) - \beta_H(s)| \leq c|t-s|^{H-\kappa}$ for all $\kappa > 0$;
- Dimension: the graph of $\beta_H(t)$ has both Hausdorff dimension and box dimension $= 2 - H$, with probability 1.

In Chapter 1 we have briefly discussed the relation between stochastic differential equations, Langevin approach, Brownian motion. Let us recall the most simple form introduced there, as described by (1.1). We assume here that $F(x, \zeta) = 0$ and $g(x, \zeta) = 1$, the resulting equation being

$$\dot{x}(t) = \xi(t), \tag{10.1}$$

where, as usual, $\xi(t)$ is a Gaussian, δ-correlated noise

$$\langle \xi(t_2)\xi(t_1) \rangle = \delta(t_2 - t_1).$$

However, this equation could be rewritten as

$$x(t) = x_0 + \int_0^t \xi(t')dt'. \tag{10.2}$$

The form (10.2) can be generalized in order to represent the fBm as (Mandelbrot and van Ness (1968); Levy (1953))

$$\begin{aligned} x^H(t) &= x_0^H + \int_0^t K(t-t')\xi(t')dt' \\ &= x_0^H + \frac{1}{\Gamma(H+\frac{1}{2})} \int_0^t [t-t']^{H-\frac{1}{2}} \xi(t')dt', \end{aligned} \tag{10.3}$$

where $\Gamma(s)$ is the Gamma function, and $\xi(t)$ is again the uncorrelated Gaussian white noise. Although there are other possible representations, we will use this form as the most adequate to our objectives, without spending time in discussing details of fractional calculus (see for instance Oldham and Spanier (1974); Podlubny (1998)). However, in the Appendix, we have indicated a few definitions.

Before we start discussing the path integral scheme we recall one of those definitions (for the *Riemann–Liouville fractional integral operator* of order $H - \frac{1}{2}$), through

$$ {}_0D_t^{-(H-\frac{1}{2})}\xi(t) = \frac{1}{\Gamma(H+\frac{1}{2})} \int_0^t [t-t']^{H-\frac{1}{2}}\xi(t')dt'. \tag{10.4} $$

This expression was initially used in a somehow cumbersome approach to the path integral representation of fBm (Sebastian (1995)). Here we will exploit this same form, but using an alternative and more direct scheme, as described in Calvo and Sánchez (2008), which is similar to the ideas used in Chapter 2.

10.2 Fractional Brownian Motion: A Path Integral Approach

As was indicated in previous chapters, given a Langevin equation describing the motion of a particle, the propagator $G(x_T, T|x_0, 0)$ is defined as the probability of finding the particle at $x = x_T$ at time $t = T$; if initially, $t = 0$, it was located at $x = x_0$. Hence, and in agreement with previous results, the propagator can be viewed as the motion of the particle averaged over all realizations of the noise compatible with the boundary conditions $x(0) = x_0$ and $x(T) = x_T$.

The essential object here is the probability measure $P(\xi(t))\mathcal{D}\xi(t)$ for the process $\xi(t)$. As was discussed several times before, since $\xi(t)$ is δ-correlated in time and distributed as a Gaussian for each t, such a probability measure is given by

$$ P(\xi(t))\mathcal{D}\xi(t) = \exp\left(-\frac{1}{2}\int_0^t \xi(s)^2 ds\right)\mathcal{D}\xi(t). \tag{10.5} $$

We want to compute the propagator of (10.3). To obtain it let us note that the boundary conditions $x(0) = x_0$, $x(T) = x_T$ can be cast into the following constraint on $\xi(t)$ (in what follows we write all expressions in terms of the exponent $\alpha = H + \frac{1}{2}$)

$$ {}_0D_t^{-(\alpha-1)}\xi(t) = x_T - x_0. \tag{10.6} $$

Therefore, the propagator can be written as the expectation value

$$ G(x_T,T|x_0,0) = \int \delta\left({}_0D_t^{-(\alpha-1)}\xi(t) - (x_T - x_0)\right) \exp\left(-\frac{1}{2}\int_0^T \xi(s)^2 ds\right)\mathcal{D}\xi(t) \tag{10.7} $$

where the Dirac delta function ensures that we only integrate over trajectories $\xi(t)$ satisfying (10.6).

We proceed now to compute the path integral on the right-hand side of (10.7). The first step is to define the action

$$ S[\xi(t)] = \frac{1}{2}\int_0^T \xi(s)^2 ds, \tag{10.8} $$

and as we have done before, consider infinitesimal variations of $\xi(t)$ satisfying (10.6). We denote by $\xi_{st}(t)$ the trajectory that makes the action *stationary* under such variations. It is worth to observe that $\xi_{st}(t)$ itself verifies (10.6). We now perform the following change of variables in (10.7)

$$ \xi(t) = \xi_{st}(t) + \eta(t). $$

The constraint given by (10.6) implies that

$$ {}_0D_t^{-(\alpha-1)}\eta(t) = 0. \tag{10.9} $$

Using that the action is quadratic in $\xi(t)$, and that $\xi_{st}(t)$ makes it stationary, we obtain the typical "semiclassical-like" result

$$ \begin{aligned} G(x_T,T|x_0,0) &= e^{-S_\alpha[\xi_{st}(t)]}\int \delta\left({}_0D_t^{-(\alpha-1)}\eta\right) e^{S[\eta(t)]}\mathcal{D}\eta(t) \\ &= F(T)e^{-S_\alpha[\xi_{st}(t)]}. \end{aligned} \tag{10.10} $$

The function $F(T)$ $(= \int \delta\left({}_0D_t^{-(\alpha-1)}\eta\right) e^{S[\eta(t)]}\mathcal{D}\eta(t))$ could be obtained at the end by imposing the normalization condition

$$ \int_{-\infty}^{\infty} G(x_T,T|x_0,0)\, dx_T = 1, \quad \forall T. \tag{10.11} $$

What remains is to compute the stationary points of $S_\alpha[\xi(t)]$ subject to the constraint (10.6). It is difficult to do it directly, because the condition (10.6) is hard to implement. However, the technique of *Lagrange multipliers* could help us. Our problem is equivalent to finding the stationary points of

$$ S_\alpha[\xi_{st}(t),\lambda] = \frac{1}{2}\int_0^T \xi(s)^2 ds + \lambda\left({}_0D_t^{-(\alpha-1)}\xi_{st}(t) - x_T + x_0\right), \tag{10.12} $$

under arbitrary infinitesimal variations of $\xi(t)$ and the Lagrange multiplier λ. From variations with respect to λ

$$\delta_\lambda S_\alpha[\xi_{st}(t), \lambda] = \delta\lambda({}_0D_t^{-(\alpha-1)}\xi_{st}(t) - (x_T - x_0)), \tag{10.13}$$

and, of course, we recover the constraint indicated in (10.6).

Let us now perform the variation respect to $\xi_{st}(t)$

$$\delta_\xi S_\alpha[\xi_{st}(t), \lambda] = \int_0^T \xi_{st}(s)\delta\xi ds + \lambda_0 D_t^{-(\alpha-1)}\delta\xi. \tag{10.14}$$

Writing explicitly the fractional integral

$$\begin{aligned}\delta\xi S_\alpha[\xi_{st}(t), \lambda] &= \int_0^T \xi_{st}(s)\delta\xi ds + \lambda\frac{1}{\Gamma(\alpha)}\int_0^T [T - t']^{\alpha-1}\delta\xi dt' \\ &= \int_0^T \left(\xi_{st}(s)ds + \lambda\frac{1}{\Gamma(\alpha)}[T - s]^{\alpha-1}\right)\delta\xi ds,\end{aligned} \tag{10.15}$$

Since $\delta_\xi S_\alpha[\xi_{st}(t), \lambda]$ must vanish for an arbitrary $\delta\xi$, we have

$$\xi_{st}(s) + \lambda\frac{1}{\Gamma(\alpha)}[T - s]^{\alpha-1} = 0. \tag{10.16}$$

The Lagrange multiplier is determined by applying ${}_0D_t^{-(\alpha-1)}$ to (10.16) and using (10.6)

$$\begin{aligned}x_T - x_0 + \lambda\frac{1}{\Gamma(\alpha)^2}\int_0^T [T - s]^{2(\alpha-1)}ds &= x_T - x_0 + \frac{\lambda}{\Gamma(\alpha)^2}\frac{T^{2\alpha-1}}{(2\alpha - 1)} \\ &\equiv 0,\end{aligned} \tag{10.17}$$

yielding

$$\lambda = -(2\alpha - 1)\Gamma(\alpha)^2\frac{x_T - x_0}{T^{2\alpha-1}}. \tag{10.18}$$

Now, inserting this value of λ into (10.16) we get the stationary trajectory

$$\xi_{st}(s) = (2\alpha - 1)\Gamma(\alpha)\frac{x_T - x_0}{T^{2\alpha-1}}[T - s]^{\alpha-1}. \tag{10.19}$$

Going back to (10.10) we straightforwardly obtain

$$G(x_T, T|x_0, 0) = F(T)\exp\left[-(\alpha - \frac{1}{2})\Gamma[\alpha]^2\frac{(x_T - x_0)^2}{T^{2\alpha-1}}\right]. \tag{10.20}$$

As indicated before, $F(T)$ is obtained from the normalization condition for $G(x_T, T|x_0, 0)$, equation (10.11), that yields

$$F(T) = \sqrt{\frac{(\alpha - \frac{1}{2})}{\pi}\frac{\Gamma(\alpha)}{T^{\alpha-\frac{1}{2}}}}. \tag{10.21}$$

Finally, the complete propagator for fBm is given by

$$\begin{aligned}G(x_T, T|x_0, 0) &= \sqrt{\frac{(\alpha - \frac{1}{2})}{\pi} \frac{\Gamma[\alpha]}{T^{\alpha-\frac{1}{2}}}} \exp\left[-(\alpha - \frac{1}{2})\Gamma[\alpha]^2 \frac{(x_T - x_0)^2}{T^{2\alpha-1}}\right].\\ &= \sqrt{\frac{H}{\pi} \frac{\Gamma(H + \frac{1}{2})}{T^H}} \exp\left[-H\Gamma(H + \frac{1}{2})^2 \frac{(x_T - x_0)^2}{T^{2H}}\right].\end{aligned} \tag{10.22}$$

If $\alpha = 1$ (or $H = \frac{1}{2}$) we get the simple diffusive case.

10.3 Fractional Brownian Motion: The Kinetic Equation

Once the propagator is known, we can easily derive the kinetic equation associated to fBm. We start considering $\hat{G}(k, T)$, the Fourier transform of $G(x, t|x_0, 0)$ respect to x

$$\hat{G}(k, T) = \exp\left[-\frac{T^{2H}}{4H\Gamma(H + \frac{1}{2})^2} k^2\right]. \tag{10.23}$$

Differentiating it respect to t, gives

$$\frac{\partial}{\partial T}\hat{G}(k, T) = -\frac{T^{2H}}{2H\Gamma(H + \frac{1}{2})^2} k^2 \hat{G}(k, T), \tag{10.24}$$

and making the inverse Fourier transform of the last equation we get the kinetic equation

$$\frac{\partial}{\partial t} G(x, t|x_0, 0) = \frac{t^{2H}}{2H\Gamma(H + \frac{1}{2})^2} \frac{\partial^2}{\partial x^2} G(x, t|x_0, 0), \tag{10.25}$$

which is a diffusion equation with a time-dependent diffusivity, as originally deduced in Wang and Lung (1990) from arguments based on the fluctuation–dissipation theorem. It is immediate to prove the invariance of (10.24) under the transformation $(x, t) \to (\mu^{\alpha-\frac{1}{2}} x, \mu t)$.

Finally, it is obvious that, when $\alpha = 1$ (i.e. $H = 1/2$), we recover the standard diffusion equation associated to Brownian motion

$$\frac{\partial}{\partial t} G(x, t|x_0, 0) = \frac{1}{2} \frac{\partial^2}{\partial x^2} G(x, t|x_0, 0). \tag{10.26}$$

10.4 Fractional Brownian Motion: Some Extensions

When considering possible extensions, it is immediate to think in the Ornstein–Uhlenbeck process (OU) (van Kampen (2004)) that was discussed in previous chapters (see for instance Chapter 7). The usual form is

$$\dot{q}(t) = -\omega q + \xi(t) \tag{10.27}$$

However, as was discussed by Eab and Lim (2006), there are in principle different possibilities for the *fractional version* of the OU process. We will analyze here a couple of those possibilities. Following Eab and Lim (2006), we could consider the following extensions (assuming $a = 0$ in the expressions of (Eab and Lim (2006)))

- Case 1: $({}_0D_t^\alpha + \omega)\, q(t) = \xi(t)$;
- Case 2: $({}_0D_t + \omega)^\alpha\, q(t) = \xi(t)$.

In what follows, and in order to simplify notation, we assume $q(t = 0) = q_o = 0$.

10.4.1 *Case 1*

Our first case is

$$({}_0D_t^\alpha + \omega)\, q(t) = \xi(t), \tag{10.28}$$

that could be immediately transformed into

$$q(t) = {}_0D_t^{-\alpha}\big(\xi(t) + \omega q(t)\big). \tag{10.29}$$

According to the procedure indicated before, for the propagator we have (remember the distribution probability for $\xi(t)$ indicated in Eq. (10.8))

$$G(q_T, T|0,0) = \int \delta\big[{}_0D_t^{-\alpha}\xi(t) - \big(q_T - \omega\, {}_0D_t^{-\alpha}q(t)\big)\big] \exp\left(-\frac{1}{2}\int_0^T \xi(s)^2 ds\right)\mathcal{D}\xi(t), \tag{10.30}$$

and as before we propose the change of variables

$$\xi(t) = \xi_{st}(t) + \eta(t),$$

where $\xi_{st}(t)$ fulfills the constraint given by (10.29), and

$${}_0D_t^{-\alpha}\eta(t) = 0.$$

We can write the propagator as indicated in (10.10), with $F(T)$ obtained from the normalization condition in (10.11). More, as the form of $F(T)$ is the same as before, we will have the the same result indicated in (10.21), that is

$$F(T) = \sqrt{\frac{\alpha - \frac{1}{2}}{\pi} \frac{\Gamma(\alpha)}{T^{\alpha - \frac{1}{2}}}}.$$

The form of the action $S_\alpha[\xi_{st}(t), \lambda]$ is

$$S_\alpha[\xi_{st}(t), \lambda] = \frac{1}{2} \int_0^T \xi(s)^2 ds + \lambda \left({}_0D_t^{-\alpha} \xi_{st}(t) - \left(q_T + \omega \, {}_0D_t^{-\alpha} q(t) \right) \right), \tag{10.31}$$

Repeating the same procedure as before (that is through variations respect to ξ_{st} and λ), we get

$$\xi_{st}(s) + \frac{\lambda (T-s)^{\alpha-1}}{\Gamma(\alpha)} \equiv 0,$$

and finally

$$\xi_{st}(t) = \frac{(T-t)^\alpha}{T^{2\alpha-1}} (2\alpha - 1) \Gamma(\alpha) \Big(q_T - \frac{\alpha}{\Gamma(\alpha)} \int_0^T (T-s)^{\alpha-1} q(s) ds \Big). \tag{10.32}$$

Hence, we obtain for the action

$$\begin{aligned} S_\alpha[\xi_{st}(t)] = \frac{(\alpha - \frac{1}{2})}{T^{2\alpha-1}} \Bigg[& \Gamma(\alpha) q_T^2 - 2\alpha q_T \int_0^T (T-s)^{\alpha-1} q(s) ds \\ & + \alpha^2 \left(\int_0^T (T-s)^{\alpha-1} q(s) ds \right)^2 \Bigg], \end{aligned} \tag{10.33}$$

and the final form of the propagator will be

$$G(q_T, T|0,0) = \sqrt{\frac{\alpha - \frac{1}{2}}{\pi} \frac{\Gamma(\alpha)}{T^{\alpha \frac{1}{2}}}} e^{-S_\alpha[\xi_{st}(t)]}. \tag{10.34}$$

If $\alpha = 1$ (or $H = \frac{1}{2}$), the last expression reduces to

$$G(q_T, T|0,0) = \frac{1}{\sqrt{2\pi T}} \exp \left(-\frac{1}{2T} \left[q_T - \int_0^T q(s) ds \right]^2 \right), \tag{10.35}$$

that corresponds to a *normal* diffusive process with (constant) absorption. Hence, we can conclude that the propagator in (10.34) would correspond to its fractional diffusive extension.

10.4.2 *Case 2*

In this case (adopting, as indicated before, $a = 0$ in the expressions of (Eab and Lim (2006))) we start from

$$({}_0D_t + \omega)^\alpha \, q(t) = \xi(t). \tag{10.36}$$

In (Eab and Lim (2006)) it was shown that the operator $({}_0D_t + \omega)^\alpha$, fulfills the relation

$$({}_0D_t + \omega)^\alpha \equiv e^{-\omega t} \, {}_0D_t^\alpha e^{\omega t},$$

allowing us to write

$$q(t) = e^{-\omega t} \, {}_0D_t^{-\alpha} e^{\omega t} \xi(t). \tag{10.37}$$

Repeating the procedure employed before, we arrive at

$$S_\alpha[\xi_{st}(t),\lambda] = \frac{1}{2}\int_0^T \xi_{st}(s)^2 ds + \lambda\left(\frac{e^{-\omega T}}{\Gamma(\alpha)}\int_0^T (T-s)^{\alpha-1} e^{\omega s}\xi_{st}(s)ds - q_T\right), \tag{10.38}$$

that yields

$$\xi_{st}(s) = q_T\Gamma(\alpha)\frac{e^{\omega T}(T-s)^{\alpha-1}e^{\omega s}}{\int_0^T (T-s)^{2(\alpha-1)}e^{2\omega s}ds}. \tag{10.39}$$

The final result for the action is

$$S_\alpha[\xi_{st}] = \frac{\Gamma(\alpha)^2}{2}\frac{q_T^2 e^{\omega T}}{\int_0^T (T-s)^{2(\alpha-1)}e^{2\omega s}ds}. \tag{10.40}$$

When $\alpha = 1$ (or $H = \frac{1}{2}$), the last expression reduces to

$$S_1[\xi_{st}] = \omega \frac{q_T^2}{(1 - e^{-2\omega T})},$$

which is the correct exponent for the *normal* OU process. Clearly, the normalization condition will give us the correct prefactor. Hence, this last case corresponds to the correct generalization of the OU process within a fBm framework.

10.5 Fractional Lévy Motion: Path Integral Approach

The case of *fractional Lévy motion* (fLm) was also studied in the literature (Calvo and Sánchez (2009); Jumarie (2007)). However, the treatment used in Jumarie (2007) is too cumbersome, while that used in Calvo and Sánchez (2009) requires to resort to properties of the Lévy distribution. For this reason we will not discuss them here, but instead we will consider a more direct (simple and probably more general) approach.

The Langevin equation defining the fLm is (Calvo and Sánchez (2009))

$$x(t) = x_o + {}_0D_t^{-(H-\frac{1}{\alpha}+1)}\eta_\alpha(t), \tag{10.41}$$

where $\eta_\alpha(t)$ is, for each t, time uncorrelated and distributed according to a symmetric Lévy distribution (Levy (1953)). The characteristic function of a symmetric Lévy distribution $L_{\alpha,\sigma}(u)$ is

$$\mathcal{F}[L_{\alpha,\sigma}](k) = \exp\left[-\sigma^\alpha |k|^\alpha\right]. \tag{10.42}$$

For $\alpha \in (0,2)$ the Lévy distribution has algebraic tails, and for $\alpha = 2$ corresponds to a Gaussian distribution. The case $\alpha = 1$ corresponds to a Lorentz distribution that is coincident with the non Gaussian process, as discussed in Chapter 8, with $q = 2$ (Prato and Tsallis (1999)). A detailed derivation of the Fourier inversion of (10.42) is given in Eliazer and Shlesinger (2012a,b).

We will consider (10.41), and in order to simply we write $H - \frac{1}{\alpha} + 1 = \nu$, that is

$$x(t) = x_o + {}_0D_t^{-\nu}\eta_\alpha(t). \tag{10.43}$$

We assume that we know $\mathcal{P}(\eta(t))$, the distribution function for the process $\eta(t)$ (where, in order to simplify notation, we eliminate the subindex α). Hence, we could write the propagator as

$$G(x_T, T|x_0, 0) = \int \mathcal{D}[\eta(t)]\delta\left[{}_0D_t^{-\nu}\eta(t) - (x_T - x_0)\right]\mathcal{P}(\eta(t)). \tag{10.44}$$

The *characteristic functional* of process η is defined through

$$\mathcal{F}[L_{\alpha,\sigma}](k(t)) = \int \mathcal{D}[\eta(t)] \exp\left[i\int_0^T ds k(s)\eta(s)\right]\mathcal{P}(\eta(t)), \tag{10.45}$$

and inverting it we get

$$\mathcal{P}(\eta(t)) = \int \mathcal{D}[k(t)] \exp\left[-\int_0^T ds\,(i\,k(s)\eta(s) + \sigma^\alpha |k(s)|^\alpha)\right]. \tag{10.46}$$

Replacing (10.46) into (10.44) we find

$$G(x_T, T|x_0, 0) = \int \mathcal{D}[\eta(t)] \int \mathcal{D}[k(t)]\ \delta\left[{}_0D_t^{-\nu}\eta(t) - (x_T - x_0)\right] \exp\left[-\int_0^T ds\,(i\,k(s)\eta(s) + \sigma^\alpha |k(s)|^\alpha)\right], \tag{10.47}$$

that is analogous to a phase space representation. Using the Fourier representation for the $\delta[.]$, the propagator adopts the form

$$\begin{aligned} G(x_T, T|x_0, 0) &= \int \mathcal{D}[\eta(t)] \int \mathcal{D}[k(t)] \int \frac{dq}{\sqrt{2\pi}} \\ &\quad \exp\left[iq\left(\int_o^T ds \frac{(T-s)^{\nu-1}}{\Gamma(\nu)}\eta(s) - (x_T - x_0)\right)\right] \\ &\quad \exp\left[-\int_0^T ds\,(i\,k(s)\eta(s) + \sigma^\alpha |k(s)|^\alpha)\right], \\ &= \int \mathcal{D}[\eta(t)] \int \mathcal{D}[k(t)] \int \frac{dq}{\sqrt{2\pi}} \\ &\quad \exp\left[i\int_o^T ds\left(q\frac{(T-s)^{\nu-1}}{\Gamma(\nu)} - k(s)\right)\eta(s)\right], \\ &\quad \exp\left[-iq(x_T - x_0) - \int_0^T ds\sigma^\alpha |k(s)|^\alpha\right]. \end{aligned} \tag{10.48}$$

The integration over $\mathcal{D}[\eta(t)]$ is now easily done as it corresponds, considering a time sliced representation, to

$$\Pi_j\ \delta\left[q\frac{(T-s_j)^{\nu-1}}{\Gamma(\nu)} - k(s_j)\right], \tag{10.49}$$

where each s_j is one of the discrete times, implying that

$$k(s) = q\frac{(T-s)^{\nu-1}}{\Gamma(\nu)} \quad \forall s.$$

We will indicate such a product by

$$\hat{\delta}\left[q\frac{(T-s)^{\nu-1}}{\Gamma(\nu)}-k(s)\right].$$

The propagator now reduces to

$$G(x_T,T|x_0,0)=\int\mathcal{D}[k(t)]\int\frac{dq}{\sqrt{2\pi}}\,\hat{\delta}\left[q\frac{(T-s)^{\nu-1}}{\Gamma(\nu)}-k(s)\right]$$
$$\exp\left[-iq(x_T-x_0)-\int_0^T ds\sigma^\alpha|k(s)|^\alpha\right].\tag{10.50}$$

Exploiting the $\delta's$ indicated in (10.49), the integration over $\mathcal{D}[k(t)]$ can be readily done. The result –now in terms of H– is

$$G(x_T,T|x_0,0)=\int_{-\infty}^{\infty}\frac{dq}{\sqrt{2\pi}}\exp\left[-iq(x_T-x_0)\right]$$
$$\exp\left[-\left(\frac{q\,\sigma}{\Gamma(H-\frac{1}{\alpha}+1)}\right)^\alpha\int_0^T ds\,(T-s)^{\alpha H-1}\right],\tag{10.51}$$

where the integral that remains in the exponential results

$$\int_0^T ds(T-s)^{\alpha H-1}=\frac{T^{\alpha H}}{\alpha H}.\tag{10.52}$$

The propagator finally takes the form

$$G(x_T,T|x_0,0)=\int_{-\infty}^{\infty}\frac{dq}{\sqrt{2\pi}}\,e^{[-iq(x_T-x_0)]}$$
$$\exp\left[-\left(\frac{q\,\sigma}{\Gamma(H-\frac{1}{\alpha}+1)}\right)^\alpha\frac{T^{\alpha H}}{\alpha H}\right].\tag{10.53}$$

10.5.1 *Gaussian Test*

There is a case where this expression could be checked: when $\alpha=2$, that is the Gaussian case. The propagator reduces to

$$G(x_T,T|x_0,0)=\int_{-\infty}^{\infty}\frac{dq}{\sqrt{2\pi}}\,e^{[-iq(x_T-x_0)]}$$
$$\exp\left[-\left(\frac{\sigma}{\Gamma(H+\frac{1}{2})}\right)^2\frac{T^{2H}}{2H}\,q^2\right].\tag{10.54}$$

This inverse Fourier transform gives the same result indicated in (10.22) for fBm. More, if $H = \frac{1}{2}$ it reduces to the simple Brownian case.

We can also look into another interesting case: $\alpha = 1$, corresponding to the Lorentz distribution indicated just before (10.43). We have

$$G(x_T, T|x_0, 0) = \int_{-\infty}^{\infty} \frac{dq}{\sqrt{2\pi}} e^{[-iq(x_T - x_0)]} \exp\left[-\left(\frac{\sigma}{\Gamma(H)}\right) \frac{T^H}{H+\frac{1}{2}} q\right].$$

$$= \frac{1}{\pi} \frac{\tilde{\sigma}}{\tilde{\sigma}^2 + (x_T - x_0)^2}, \tag{10.55}$$

where

$$\tilde{\sigma} = \frac{\sigma}{(H+\frac{1}{2})\,\Gamma(H)} T^H.$$

10.5.2 *Kinetic Equation*

We can applied the same procedure used for the case of fBm in order to obtain the associated kinetic equation (Calvo and Sánchez (2009)). Consider the Fourier transform of (10.53)

$$G(q, T) = \exp\left[-\left(\frac{q\,\sigma}{\Gamma(H - \frac{1}{\alpha} + 1)}\right)^\alpha \frac{T^{\alpha H}}{\alpha H}\right]. \tag{10.56}$$

Differentiating it with respect to T we have

$$\frac{\partial}{\partial T} G(q, T) = -\left(\frac{q\,\sigma}{\Gamma(H - \frac{1}{\alpha} + 1)}\right)^\alpha T^{\alpha H - 1} G(q, T). \tag{10.57}$$

Making the inverse Fourier transform and taking into account the identity

$$\mathcal{F}\left[\frac{\partial^\alpha f}{\partial |x|^\alpha}\right] = -|q|^\alpha f(q)$$

where $\mathcal{F}[.]$ indicates the Fourier transform, we obtain

$$\frac{\partial}{\partial T} G(x, T) = \left(\frac{\sigma}{\Gamma(H - \frac{1}{\alpha} + 1)}\right)^\alpha T^{\alpha H - 1} \frac{\partial^\alpha}{\partial |x|^\alpha} G(x, T), \tag{10.58}$$

indicating that the fLm propagator fulfills a space-fractional diffusion-like equation with a time dependent diffusivity

For the Gaussian case ($\alpha = 2$ and $H = \frac{1}{2}$) the form (10.58) reduces to

$$\frac{\partial}{\partial T} G(x, T) = \frac{\sigma^2}{2} \frac{\partial^2}{\partial x^2} G(x, T),$$

the usual diffusion case, while for the Lorentz case ($\alpha = 1$) we get

$$\frac{\partial}{\partial T} G(x, T) = \frac{\sigma T^{H-1}}{\Gamma(H)} \frac{\partial}{\partial |x|} G(x, T).$$

10.6 Fractional Lévy Motion: Final Comments

The result we have shown in (10.53) completely agrees with the one found by Calvo and Sánchez (2009). The main advantage of our procedure relies in the fact that is less restrictive as it does not require to exploit properties of Lévy distributions. Besides, it could be readily applied to other different distributions.

More, as indicated in (Calvo and Sánchez (2009)), this result implies that the propagator for fLm is a Lévy distribution that depends on the relation $\Delta x/T^H$, with the average motion being self-similar with an exponent H.

Let us also sketch the possible extensions as done before for fBm. We start considering the analogous of the first case

$$x(t) = {}_0D_t^{-\nu}(\eta_\alpha + \omega x(t)), \tag{10.59}$$

whose propagator will have the form

$$\begin{aligned} G(x_T, T|x_0, 0) = \int \mathcal{D}[\eta(t)] \int \mathcal{D}[k(t)] \int \frac{dq}{\sqrt{2\pi}} \\ \exp\left[iq\left(\int_o^T ds \frac{(T-s)^{\nu-1}}{\Gamma(\nu)}(\eta(s)+x(s)) - x_T\right)\right] \\ \exp\left[-\int_0^T ds\,(i\,k(s)\eta(s) + \sigma^\alpha|k(s)|^\alpha)\right]. \end{aligned} \tag{10.60}$$

After applying the same procedure as before we arrive at

$$\begin{aligned} G(x_T, T|x_0, 0) = \int_{-\infty}^{\infty} \frac{dq}{\sqrt{2\pi}} \exp\left[-iq(x_T - \int_0^T ds \frac{(T-s)^{H-\frac{1}{\alpha}}}{\Gamma(\nu)} x(s))\right] \\ \exp\left[-\left(\frac{q\,\sigma}{\Gamma(H-\frac{1}{\alpha}+1)}\right)^\alpha \frac{T^{\alpha H}}{\alpha H}\right]. \end{aligned} \tag{10.61}$$

This last expression, when $\alpha = 2$ and $H = \frac{1}{2}$ reduces to the known diffusion process with absorption. Hence, the last expression corresponds to a Lévy-driving diffusion with absorption.

For the second case we again resort to (Eab and Lim (2006))) starting from

$$({}_0D_t + \omega)^\nu\, x(t) = \eta_\alpha(t), \tag{10.62}$$

and using the relation indicated after (10.36)

$$({}_0D_t + \omega)^\nu \equiv e^{-\omega t}\,{}_0D_t^\nu e^{\omega t},$$

we can write (without writing the subindex α)

$$x(t) = e^{-\omega t}\,{}_0D_t^{-\nu} e^{\omega t}\eta(t). \tag{10.63}$$

The form of the propagator will be

$$\begin{aligned} G(x_T, T|x_0, 0) = & \int \mathcal{D}[\eta(t)] \int \mathcal{D}[k(t)] \int \frac{dq}{\sqrt{2\pi}} \\ & \exp\left[iq\left(e^{-\omega T}\int_o^T ds \frac{(T-s)^{\nu-1}}{\Gamma(\nu)} e^{-\omega s}\eta(s) - x_T\right)\right] \\ & \exp\left[-\int_0^T ds\,(i\,k(s)\eta(s) + \sigma^\alpha |k(s)|^\alpha)\right]. \end{aligned} \tag{10.64}$$

Repeating the procedure employed before, we arrive at

$$\begin{aligned} G(x_T, T|x_0, 0) = & \int_{-\infty}^{\infty} \frac{dq}{\sqrt{2\pi}} \exp\left[-iq\left(x_T - \frac{e^{-\omega T}}{\Gamma(\nu)}\int_0^T ds (T-s)^{\nu-1} e^{\omega s}\right)\right] \\ & \exp\left[-\left(\frac{q\,\sigma}{\Gamma(\nu)}\right)^\alpha e^{-\alpha\omega T}\int_0^T ds (T-s)^{\alpha(\nu-1)} e^{\alpha\omega s}\right]. \end{aligned} \tag{10.65}$$

When $\alpha = 2$ and $H = \frac{1}{2}$ the last expression reduces to the well known propagator form of the OU process (in fact its Fourier transform). The last expression will hence correspond to the fLm generalization of the OU process.

We stop here and do not pursue this analysis further. We left to complete the details of the last two examples as exercises for the diligent reader!

Chapter 11

Feynman–Kac Formula, the Influence Functional

11.1 Feynman–Kac formula

The Feynman–Kac formula (Kac (1949, 1959); Feynman and Hibbs (1965)), named after Richard Feynman and Mark Kac, establishes a link between parabolic partial differential equations (PDEs) and stochastic processes. It offers a method of solving certain PDEs by simulating random paths of a stochastic process. Conversely, an important class of expectations of random processes can be computed by deterministic methods. Consider the PDE,

$$\frac{\partial\varphi(x,t)}{\partial t} + \mu(x,t)\frac{\partial\varphi(x,t)}{\partial x} + \frac{\sigma(x,t)}{2}\frac{\partial^2\varphi(x,t)}{\partial x^2} = V(x,t)\varphi(x,t), \quad (11.1)$$

with the final condition $\varphi(x,t=t_f) = \psi(x)$, where $\mu(x,t)$, $\sigma(x,t)$ and $V(x,t)$ are known functions and $\varphi(x,t)$ is the unknown. Feynman–Kac formula tell us that the solution could be written as an *expectation*

$$\varphi(x,t) = \mathcal{E}\Big[\exp\Big(\int_{t_o}^{t_f} V(z,\tau)d\tau\Big)\psi(z)\Big|z=x\Big], \quad (11.2)$$

where z is a stochastic process driven by

$$dz = \mu(z,t)dt + \sigma(z,t)^{1/2}dW, \quad (11.3)$$

with dW the Wiener process, and the initial condition is $z(0) = x$. This expectation can be estimated by Monte Carlo or other numerical methods (Kwas (2005)).

However, the Feynman–Kac formula can be also used to look at the *evolution* of a given initial condition by an equation like (11.1). Let us look at the derivation of such a formula, following a procedure similar to the one used by Kac (Kac (1949, 1959); Schulman (1981)).

Let $\mathcal{Q}$ be a function of the path $q(t)$. Its expectation could be written as

$$\mathcal{E}[\mathcal{Q}] = \sum_q q P_W[q(t)) \in \Omega | \mathcal{Q}[q(t)] = q], \tag{11.4}$$

where Ω is the space of all path starting at $(q(t_o) = q_o, t = t_o)$ and reaching $(q(t_f) = q_f, t = t_f)$; $P_W[q(t)) \in \Omega | \mathcal{Q}[q(t)] = q]$ is the non normalized Wiener measure of the set of path on which $\mathcal{Q}[q(t)]$ takes the value q. This expectation can be also written as

$$\begin{aligned}
\mathcal{E}[\mathcal{Q}] &= \lim_{N\to\infty} \int_{-\infty}^{\infty} dq_1 \int_{-\infty}^{\infty} dq_2 \ldots \int_{-\infty}^{\infty} dq_{N-1} P(q_1, t_1) \\
&\quad \times P(q_2, t_1 | q_1, t_1) \ldots P(q_f, t_f | q_{N-1}, t_{N-1}) \mathcal{Q}[q_1, q_2, \ldots, q_{N-1}], \\
&= \lim_{N\to\infty} \int_{-\infty}^{\infty} dq_1 \int_{-\infty}^{\infty} dq_2 \ldots \int_{-\infty}^{\infty} dq_{N-1} P(q_1, t_1) \\
&\quad \times P(q_2 - q_2, t_2 - t_1) \ldots P(q_f - q_{N-1}, t_f - t_{N-1}) \\
&\quad \mathcal{Q}[q_1, q_2, \ldots, q_{N-1}],
\end{aligned} \tag{11.5}$$

$\mathcal{Q}[q_1, q_2, \ldots, q_{N-1}]$ is the function evaluated along the trajectory, and in the last line we have used translational invariance. This is not normalized as we have not integrated over q_N. According to what we know for the Wiener process

$$\mathcal{E}[1] = \frac{1}{\sqrt{4\pi D t_f}} \exp\left[-\frac{q_f^2}{4Dt_f}\right]. \tag{11.6}$$

In Chapter 2 we have seen how it was possible to recover the Fokker–Plank equation from the path integral form of the propagator. In other words, we proved that the propagator's path integral representation is solution of the Fokker–Plank equation. Following Kac (Kac (1949, 1959); Schulman (1981)) we now show that the functional

$$\phi(q,t) = \mathcal{E}\left[\exp\left(\int_{t_o}^{t_f} U(q(\tau)) d\tau\right)\right], \tag{11.7}$$

with the initial condition $\phi(q, t = t_o) = \delta(q)$, is a solution of the PDE

$$\frac{\partial}{\partial t}\phi(q,t) = D\frac{\partial^2}{\partial q^2}\phi(q,t) - U(q)\phi(q,t).$$

It is well known the linearity of expectation values, hence (without too much rigor!)

$$\phi(q,t) = \sum_{k=0}^{\infty} \frac{(-1)^k}{k!} \mathcal{E}\left[\left(\int_{t_o}^{t_f} U(q(\tau))d\tau\right)^k\right]. \tag{11.8}$$

Let us consider the different terms in the sum. For $k = 0$ we have

$$\phi_o(q,t) = \mathcal{E}[1] = \frac{1}{\sqrt{4\pi D t_f}} \exp\left[-\frac{q_f^2}{4Dt_f}\right], \tag{11.9}$$

while for $k = 1$

$$\phi_1(q,t) = \mathcal{E}\left[\left(\int_{t_o}^{t_f} U(q(\tau))d\tau\right)\right] = \int_{t_o}^{t_f} d\tau \mathcal{E}[(U(q(\tau)))]. \tag{11.10}$$

The expected value of $U(q(\tau))$ is the probability of $q(\tau)$ taking some value ζ times $U(\zeta)$ and summed over ζ. Hence

$$\begin{aligned}\phi_1(q,t) &= \int_{t_o}^{t_f} d\tau \int_{-\infty}^{\infty} d\zeta P(q_f - \zeta, t_f - \tau) U(\zeta) P(\zeta, \tau) \\ &= \int_{t_o}^{t_f} d\tau \int_{-\infty}^{\infty} d\zeta P(q_f - \zeta, t_f - \tau) U(\zeta) \phi_0(\zeta, \tau).\end{aligned} \tag{11.11}$$

Now, for $k = 2$ it is clear that we have two integrals. It results convenient to define

$$\phi_2(q,t) \equiv \frac{1}{2}\mathcal{E}\left[\left(\int_{t_o}^{t_f} d\tau_1 U(q(\tau_1)) \int_{t_o}^{t_f} d\tau_2 U(q(\tau_2)),\right)\right], \tag{11.12}$$

and we need to invoke the *Lema*

$$\left(\int_{t_o}^{t_f} g(s)ds\right)^n = n! \int_{t_o}^{t_f} ds_1 \int_{t_o}^{t_f} ds_2 \dots \int_{t_o}^{t_f} ds_n g(s_1)g(s_2)\dots g(s_n), \tag{11.13}$$

that can be easily proved by induction. Using this result, (11.12) becomes

$$\phi_2(q,t) = \int_{t_o}^{t_f} d\tau_1 \int_{t_o}^{t_f} d\tau_2 \mathcal{E}\left[U(q(\tau_1))U(q(\tau_2))\right], \tag{11.14}$$

that similarly to (11.11) can be rearranged as

$$\begin{aligned}\phi_2(q,t) = \int_{t_o}^{t_f} d\tau_1 \int_{t_o}^{t_f} d\tau_2 \int_{-\infty}^{\infty} d\zeta_1 \int_{-\infty}^{\infty} d\zeta_2 P(q_f - \zeta_1, t_f - \tau_1) U(\zeta_1) \\ \times P(\zeta_1 - \zeta_2, \tau_1 - \tau_2) U(\zeta_2) P(\zeta_2, \tau_2) \\ = \int_{t_o}^{t_f} d\tau \int_{-\infty}^{\infty} d\zeta P(q_f - \zeta, t_f - \tau) U(\zeta) \phi_0(\zeta, \tau).\end{aligned} \tag{11.15}$$

Now, if for $k = 1, 2, ...$ we let

$$\phi_k(q,t) \equiv \frac{1}{k!}\mathcal{E}\left[\left(\int_{t_o}^{t_f} d\tau U(q(\tau))\right)^k\right], \tag{11.16}$$

it follows that

$$\phi_k(q,t) = \int_{t_o}^{t_f} d\tau \int_{-\infty}^{\infty} d\zeta P(q_f - \zeta, t_f - \tau) U(\zeta) \phi_{k-1}(\zeta, \tau). \tag{11.17}$$

Returning now to (11.7) we can write

$$\begin{aligned}
\phi(q,t) &= \sum_{k=0}^{\infty} (-1)^k \phi_k(q,t) \\
&= \phi_o(q,t) - \sum_{k=1}^{\infty} (-1)^{k-1} \int_{t_o}^{t_f} d\tau \int_{-\infty}^{\infty} d\zeta \phi_o(q_f - \zeta, t_f - \tau) \\
&\qquad \times U(\zeta)\phi_{k-1}(\zeta, \tau) \\
&= \phi_o(q,t) - \int_{t_o}^{t_f} d\tau \int_{-\infty}^{\infty} d\zeta \phi_o(q_f - \zeta, t_f - \tau) U(\zeta) \phi(\zeta, \tau).
\end{aligned} \tag{11.18}$$

Recalling that

$$\left(\frac{\partial}{\partial t} - D\frac{\partial^2}{\partial q^2}\right)\phi_o(q,t) = 0$$

$$\lim_{t \to 0} \phi_o(q,t) = \delta(q),$$

the integral equation for $\phi(q,t)$ implies that

$$\frac{\partial}{\partial t}\phi(q,t) = D\frac{\partial^2}{\partial q^2}\phi(q,t) - U(q)\phi(q,t). \tag{11.19}$$

More in general, if we have $t_o = 0$, $\phi(q,0) = v(q)$, we can now write that

$$\phi(q,t) = \int \mathcal{D}[x] v(x(t) + q) \exp\left[\int_0^t d\tau V(x(\tau) + q)\right], \tag{11.20}$$

is the solution of

$$\frac{\partial}{\partial t}\phi(q,t) = \frac{1}{2}\frac{\partial^2}{\partial q^2}\phi(q,t) + U(q)\phi(q,t). \tag{11.21}$$

This results, as extended by Freidlin (1985), was exploited in (Hassan *et al.* (1994)) to analyze front propagation in reaction diffusion systems of the form

$$\frac{\partial}{\partial t}\phi(q,t) = \frac{\partial^2}{\partial q^2}\phi(q,t) + f(\phi(q,t)). \tag{11.22}$$

If both, $f(\phi(q,t))$ and $f(\phi(q,t))/\phi(q,t)$ are bounded functions for $t_o \leq t \leq \infty$ and $-\infty \leq q \leq \infty$, hence

$$\phi(q,t) = \int \mathcal{D}[q(t)]\phi_o(q) \exp\left[-\int_{t_o}^{t} ds \left(\frac{1}{4}\dot{q}(s)^2 + \frac{f(\phi(q,s))}{\phi(q,s)}\right)\right], \tag{11.23}$$

is solution of (11.22) with initial condition $\phi(q,t_o) = \phi_o(q)$. In (Hassan *et al.* (1994)) the above indicated solution was transformed into a short-time expression more adequate for numerical studies of the sought evolution.

As a final comment, we want to indicate that there is a recent *fractional* generalization of Kac's formula due to Tarasov and Zaslavsky (2008).

11.2 Influence Functional: Elimination of Irrelevant Variables

Here we introduce an approach, based on Feynman's influence functional method (Feynman and Vernon (1963)), for eliminating irrelevant variables in stochastic processes whose complete description is Markovian. As this approach does not resort to adiabatic arguments, it allows (in principle) to eliminate either fast or slow variables (van Kampen (1985)). Here we discuss the case of multivariate Fokker–Planck equations, although the method can be applied to the equivalent set of coupled Langevin equations with (additive or multiplicative) white noise sources. We will focus the emphasis on the form of the propagator for the relevant variables instead of on the structure of the reduced evolution equation. Here we will follow the lines of the work presented in Wio (1986); Wio *et al.* (1993a, 1995).

For simplicity we will work with only two variables, the extension to many variables being tedious but straightforward. We will use the notation and several results of Langouche *et al.* (1982). We start from the Fokker–Planck equation

$$\frac{\partial P}{\partial t} = \frac{\partial}{\partial x}[h_x(x,y)\,P] + \frac{\partial}{\partial y}[h_y(x,y)\,P] + \frac{1}{2}\left[D_{xx}\frac{\partial^2}{\partial x^2} + 2D_{xy}\frac{\partial^2}{\partial x \partial y} + D_{yy}\frac{\partial^2}{\partial y^2}\right] P, \tag{11.24}$$

for the conditional probability $P(x,y,t|x_0,y_0,t_0)$ to reach (x,y) at time t starting from (x_0,y_0) at t_0. The equivalent Langevin equations correspond to uncorrelated additive white noises. The corresponding full propagator is (Langouche *et al.* (1982))

$$P(x, y, t|x_0, y_0, t_0) = \int \mathcal{D}[x]\mathcal{D}[y] \exp\left[-\int_{t_0}^{t} dt\, \mathcal{L}(x, y, \dot{x}, \dot{y})\right], \tag{11.25}$$

with $\mathcal{D}[x]\mathcal{D}[y]$ the usual measure.

Let us call x the *relevant* and y the *irrelevant* variables, and separate

$$\begin{aligned} h_x(x, y) &= h_{0,x}(x) + h_{1,x}(x, y), \\ h_y(x, y) &= h_{0,y}(y) + h_{1,y}(x, y). \end{aligned} \tag{11.26}$$

The full Lagrangian

$$\mathcal{L}(x, y, \dot{x}, \dot{y}) = \frac{1}{2}\left[\tilde{D}_{xx}(\dot{x} - h_x)^2 + \tilde{D}_{yy}(\dot{y} - h_y)^2 + 2\tilde{D}_{xy}(\dot{x} - h_x)(\dot{y} - h_y)\right], \tag{11.27}$$

where $\tilde{D}_{i,j}$ are the elements of the inverse diffusion matrix. It can then be written as the sum of three contributions,

$$\mathcal{L}(x, y, \dot{x}, \dot{y}) = \mathcal{L}_x(x, \dot{x}) + \mathcal{L}_y(y, \dot{y}) + \mathcal{L}_{\text{int}}(x, \dot{x}, y, \dot{y}). \tag{11.28}$$

Note that $\mathcal{L}_x$ and $\mathcal{L}_y$ might not be *bona fide* Lagrangian (that is with a structure like (2.18)). We define the marginal or inclusive conditional probability as

$$P_{\text{incl}}(x_f, t_f|x_0, t_0) = \int dy_f \int dy_0\, P_y(y_0) P(x_f, y_f, t_f|x_0, y_0, t_0), \tag{11.29}$$

where we have averaged over the initial distribution $P_y(y_0)$ of y [assuming the joint initial distribution to be separable: $P_0(x_0, y_0) = P_x(x_0)P_y(y_0)$], and summed over all possible final values y_f as usual. Using (11.25) this can be recast into the form

$$P_{\text{incl}}(x_f, t_f|x_0, t_0) = \int \mathcal{D}[x] \exp\left[-\int_{t_0}^{t_f} dt\, \mathcal{L}_x(x, \dot{x})\right] \mathcal{F}[x, \dot{x}], \tag{11.30}$$

$$\mathcal{F}[x, \dot{x}] = \int dy_f \int dy_0\, P_y(y_0) \int \mathcal{D}[y] \exp\left[-\int_{t_0}^{t_f} dt\, (\mathcal{L}_y + \mathcal{L}_{\text{int}})\right], \tag{11.31}$$

where $\mathcal{F}$ has the structure of Feynman's influence functional (Feynman and Vernon (1963)), although the latter is a double path integral instead. In the same spirit of Feynman's scheme, we write

$$\mathcal{F}[x, \dot{x}] = \exp\left\{-\int_{t_0}^{t_f} dt\, \Phi[x(t), \dot{x}(t)]\right\}. \tag{11.32}$$

We now introduce for $\Phi[x, \dot{x}]$ the Ansatz

$$\Phi[x(t), \dot{x}(t)] = f_0(t)x(t) - \int_{t_0}^{t_f} d\tau\, x(t)C(t,\tau)x(\tau), \tag{11.33}$$

$$f_0(t) = f_{00}(x_f, x_0)\delta(t - t_0) + \int_{t_0}^{t_f} d\tau\, f_{0,1}(t,\tau)x(\tau),$$

with $f_{00}(x_f, x_0)$ some function of x_0 and x_f, the initial and final coordinates respectively. In order to see how such a structure could arise in a practical case we will assume the simplified situation

$$\begin{aligned} h_{0,x}(x) &= Ax, & h_{1,x}(x,y) &= \lambda g(y), \\ h_{0,y}(y) &= h(y), & h_{1,y}(x,y) &= \lambda Bx, \end{aligned}$$
$$D_{ij} = D, \quad i,j = x,y. \tag{11.34}$$

The interaction Lagrangian becomes

$$\begin{aligned} \mathcal{L}_{\text{int}} &= (\lambda/D)\Big\{[Ag(y) + Bh(y) - B\dot{y}]x - g(y)\dot{x}\Big\} \\ &= \lambda[\eta(y,\dot{y})x - \varphi(y)\dot{x}], \end{aligned} \tag{11.35}$$

$\eta(y, \dot{y})$ being a function of y and $\dot{y}$, and $\varphi(y)$ of y only. Replacing (11.35) in (11.31) gives

$$\begin{aligned} \mathcal{F}[x,\dot{x}] = \int dy_f \int dy_0 P_y(y_0) \int \mathcal{D}[y] \ \exp\left[-\int dt\, \mathcal{L}_y\right] \\ \times \exp\left[-\lambda \int dt\, (\eta x - \varphi \dot{x})\right]. \end{aligned} \tag{11.36}$$

In the case of weak coupling ($\lambda \ll 1$), we can expand the second exponential in powers of the coupling constant and perform the path integral, obtaining

$$\begin{aligned} \mathcal{F}[x,\dot{x}] = \mathcal{F}_0\Big\{1 &- \lambda \int dt \Big[\langle\eta(y,\dot{y})\rangle x - \langle\varphi(y)\rangle\dot{x}\Big] \\ &+ \lambda^2 \int dt \int dt' \Big[x\langle\eta(y,\dot{y})\eta(y',\dot{y}')\rangle x' + x\langle\eta(y,\dot{y})\varphi(y')\rangle\dot{x}' \\ &+ \dot{x}\langle\varphi(y)\eta(y',\dot{y}')\rangle x' + \dot{x}\langle\varphi(y)\varphi(y')\rangle\dot{x}'\Big] + \cdots\Big\}, \end{aligned} \tag{11.37}$$

where $x = x(t)$, $x' = x(t')$, etc. The angular brackets denote the average

$$\langle\phi(y)\rangle = \mathcal{F}_0^{-1} \int dy_0 \int dy_f P_y(y_0) \int \mathcal{D}[y]\, \phi(y) \exp\left[-\int dt\, \mathcal{L}_y\right], \tag{11.38}$$

and $\mathcal{F}_0$ will be unity only if $\mathcal{L}_y$ is a "bona fide" Lagrangian (of the form (2.18)). Transforming the perturbation expansion (11.37) into a cumulant expansion, and restricting ourselves to the case in which second order perturbation theory leads to reasonable accuracy, the influence functional reduces to

$$\begin{aligned}\mathcal{F}[x,\dot{x}] = \mathcal{F}_0 \ \exp\Big\{ & -\lambda \int dt \, [x(t)\eta_0(t) - \dot{x}(t)\varphi_0(t)] \\ & + \lambda^2 \int dt \int dt' \, [x(t)\eta_1(t,t')x(t') \\ & + \cdots + \dot{x}(t)\varphi_1(t,t')\dot{x}(t')] \Big\}, \end{aligned} \tag{11.39}$$

$$\begin{aligned}\eta_0(t) &= \langle \eta(y(t),\dot{y}(t)) \rangle, & \varphi_0(t) &= \langle \varphi(y(t)) \rangle, \\ \eta_1(t,t') &= \langle \eta(y(t),\dot{y}(t))\eta(y(t'),\dot{y}(t')) \rangle, & \varphi_1(t,t') &= \langle \varphi(y(t))\varphi(y(t')) \rangle, \quad \text{etc.}\end{aligned}$$

We see that (11.39) has the desired structure (11.33) (after integrating by parts). The above Ansatz can be interpreted as the effect of an *effective* Gaussian colored noise, coming from the eliminated irrelevant variable, on the relevant one (Wio *et al.* (1995)).

11.2.1 *Example: Colored Noise*

For this first case we consider an Ornstein–Uhlenbeck noise,

$$\begin{aligned}\dot{x} &= h(x) + g(x)\varepsilon, \\ \dot{\varepsilon} &= -\tau^{-1}u + \tau^{-2}\xi(t), \end{aligned} \tag{11.40}$$

τ being the correlation time and $\xi(t)$ a Gaussian white noise of zero mean and δ correlated. This is the problem we have discussed in Chapter 7. The corresponding FPE has a singular diffusion tensor, calling for a "phase space" treatment of the path integrals (Wio *et al.* (1989)). After integration over ε and its associate conjugate variable, we get a "phase space" influence functional with the same structure as the proposed Ansatz. At this stage we can expand the exponents in powers of τ (we note that in the original problem the limit $\tau \to 0$ corresponds to white noise). However, we should keep in mind that such an expansion cut at 2nd order! Up to zero order in τ we find a marginal conditional probability corresponding to the Stratonovich prescription (i.e. the middle point discretization we have been working with) for a multiplicative white noise (Langouche *et al.* (1982)). For additive noise $[g(x) = 1]$, the next order in τ coincides with the lowest order of the "exact" result of Pesquera *et al.* (1983). We will not pursue the analysis of higher order terms.

11.2.2 *Example: Lotka–Volterra Model*

We consider now the simplified Lotka–Volterra model

$$\begin{aligned} \dot{x} &= by + \xi_1(t), \\ \dot{y} &= -bx + \xi_2(t), \end{aligned} \tag{11.41}$$

where, as before, the ξ_j are white noises with zero mean and correlation $\langle \xi_i(t)\xi_j(t') \rangle = 2D\delta_{ij}\delta(t-t')$. As the involved Lagrangian turns out to be quadratic in the coordinates and velocities, the problem can be exactly integrated. Particularly the influence functional has an exact expression with the same structure as the Ansatz (note that both variables clearly evolve on the same time scale, so an adiabatic elimination procedure is not applicable). In this case the evolution equation for the marginal conditional probability can be readily derived. It has the form of a Fokker–Planck equation but with a drift coefficient explicitly dependent on time and on the initial conditions, making evident the non-Markovian character of the reduced problem (Wio *et al.* (1995)).

11.3 Kramers Problem

The Kramers problem, already discussed in a previous chapter, could also be casted into a form similar to the previous cases, and discussed from the point of view of an influence functional. That is, looking for a Markovian approximation to this problem. However, we will not repeat the calculation here.

Chapter 12

Other Diffusion-Like Problems

In this chapter we discuss other two stochastic examples where the path integral approach has been useful. The first corresponds to a modeling of a particle moving in a media with a nonlinear shear flow, while the second one is a reaction-diffusion model for diffusion controlled reactions.

12.1 Diffusion in Shear Flows

The phenomenon of a Brownian particle moving in a media with shear flow has attracted much interest in order to describe faster than diffusive motion of tracer particles. A simple bidimensional model representation of the problem have the form

$$\frac{\partial}{\partial t}P(x,y,t|x_o,y_o,t_o) = -V(y)\frac{\partial}{\partial x}P + D\frac{\partial^2}{\partial y^2}P, \tag{12.1}$$

where $P(x,y,t|x_o,y_o,t_o)$ is the conditional probability to reach the point (x,y) at time t, starting from (x_o,y_o) at time t_o, $V(y)$ is the shear velocity field and D is the diffusion coefficient associated to the diffusive motion in the y direction.

Several studies have shown that if the velocity field has a linear dependence on y, the mean square displacement growths as $\langle x(t)^2\rangle \sim t^3$, while for random velocity fields, and within a continuous-time random-walk formalism, the exact asymptotic behavior of the first few nontrivial displacement moments were found. Such analysis provided information about the scaling of the marginal conditional probability

$$P(x,t) \propto \int dy P(x,y,t|x_o,y_o,t_o).$$

A case of much interest was when there was a power-law shear flow, that is

$$V(y) \propto sgn(y)y^\beta,$$

with $sgn(y)$ indicating the sign function. It was found that when varying the power β from 1 to 0, and for a δ-like initial condition $P(x_o, y_o, t_o) \propto \delta(x-x_o)\delta(y-y_o)$, the marginal conditional probability exhibits a transition from unimodal to bimodal behavior. The critical value was theoretically estimated to be $\beta = 9/8$, while numerically resulted $\beta = 0.75$.

Here we present a path integral approach to this problem, based on the work of Wio and Zanette (1993). As we have seen before, when the associated Lagrangian is at most quadratic y coordinates and velocities, the path integral representation of the conditional probability can be exactly integrated, for instance exploiting the path expansion method. Clearly, to analyze the problem with a power-law shear flow could be far from trivial. For this reason we adopted the following piecewise linear approximation

$$V(y) = v_o \begin{cases} \frac{\epsilon y}{a+(1-\epsilon^2)}, & \epsilon a < y, \\ \frac{y}{\epsilon a}, & -\epsilon a < y < \epsilon a, \\ \frac{\epsilon y}{a-(1-\epsilon^2)}, & y < \epsilon a, \end{cases} \tag{12.2}$$

where a and ϵ are constants $(0 \leq \epsilon \leq 1)$. For $\epsilon = 0$ and $\epsilon = 1$ the form of the velocity field coincides with the studied power-law shear flows when $\beta = 0$ and $\beta = 1$. For intermediate values of ϵ and β, it is expected to be a nice mimic of a power-law field. In addition, the indicated linearization is the first step in a systematic polygonal approximation to a power-law velocity field. The present piecewise linear approximation allows us to obtain the reference path and to perform the path integration by means of the path expansion procedure.

Using a phase-space representation, the conditional probability could be written as

$$P(x,y,t|x_o,y_o,t_o) = \int \mathcal{D}[x]\mathcal{D}[y]\mathcal{D}[p]\mathcal{D}[q] \exp\{\mathcal{S}[x,y,p,q]\}, \tag{12.3}$$

where p and q are the conjugate "momentum" of the coordinates x and y respectively. The action $\mathcal{S}[x,y,p,q]$ is given by

$$\mathcal{S}[x,y,p,q] = \int_{t_o}^{t} ds \left[ip(s)(\dot{x}(s) - V(y(s))) + iq(s)\dot{y}(s) + Dq(s)^2\right]. \tag{12.4}$$

Due to the Gaussian dependence on q, the integral over this variable is immediate. The integration over p can also be done, leading to a δ-functional, making that the integration over x can be performed. We consider the initial condition $P(x_o, y_o, t_o) \propto \delta(x_o)\delta(y_o)$ at $t_o = 0$ and simplified denoting $P(x,y,t) \equiv P(x,y,t|0,0,0)$.

Using a Fourier representation of the resulting δ-function in x and $V(y)$, we obtain

$$P(x,y,t) = \int \frac{dk}{\sqrt{2\pi}} \exp\{ik(x-x_o)\}\mathcal{P}(y,t)$$
$$\mathcal{P}(y,t) = \int \mathcal{D}[y] \exp\left\{-\int_0^t ds\left(ikV(y(s)) + \frac{\dot{y}^2}{4D}\right)\right\}. \quad (12.5)$$

From the variation of the Lagrangian-like functional in the exponent of the 2nd integral of (12.5) we get the Euler–Lagrange equation for the "most probable" trajectory as

$$\ddot{y} - iDk\frac{\partial}{\partial y}V(y) = 0, \quad (12.6)$$

that should be solved with the boundary conditions $y(s=0) = 0$ and $y(s=t) = y$. Clearly, the solutions of (12.6) correspond to complex trajectories. However, as discussed in Garrison and Wright (1985); Wio and Zanette (1993), this fact introduces no major difficulties. In the present case, the simplest way to perform the path integral is to use in (12.6) the indicated boundary conditions in order to fix the real part of the trajectory, as the imaginary part is automatically determined. Hence, we can use this trajectory as the reference path in the path-expansion scheme.

The symmetry of the problem: $P(x,y,t) = P(-x,-y,t)$, implies that it is enough to only consider the positive values of y. As a consequence of the discontinuity in the potential's derivative, the reference trajectory should be split according to the value of y being smaller or larger than ϵa. This also implies that $P(x,y,t)$ is also split into two functional forms. For $y < \epsilon a$ we have

$$P(x,y,t) = \frac{1}{\sqrt{4\pi d a^2 t}} \exp\left\{-\frac{\eta^2}{4dt}\right\}\left(\frac{12\epsilon^2}{dv_o^2t^3}\right)^{1/2} \exp\left\{-\frac{3\epsilon^2}{dv_o^2t^3}\left[x - \frac{v_o\eta t}{2\epsilon}\right]^2\right\}, \quad (12.7)$$

where we have the scaled coordinate $\eta = y/a$, and $d = \frac{D}{a^2}$. For $y > \epsilon a$ we have

$$P(x,y,t) = \frac{1}{\sqrt{4\pi da^2 t}} \exp\left\{-\frac{\eta^2}{4dt}\right\} \int \frac{dk}{\sqrt{2\pi}}$$
$$\times \left[1 + i\frac{2kv_o dt^2}{\eta^2}\left(1 - \frac{\epsilon}{\eta}(1-\epsilon^2)\right)^{\frac{1}{2}}\right]$$
$$\times \exp\left\{-k^2 \frac{dv_o^2 t^3 \epsilon}{12\eta^2}\phi(\eta,\epsilon) + ik\psi(x,\eta,\epsilon)\right\}, \quad (12.8)$$

where

$$\phi(\eta,\epsilon) = \epsilon\eta^4 + 6\epsilon(1-\epsilon^2)\eta^2 + 4(1-2\epsilon^2)(1-\epsilon^2)\eta - 3\epsilon(1-\epsilon^2)^2$$
$$\psi(x,\eta,\epsilon) = x - v_o t\eta\left(\frac{\epsilon}{2} + (1-\epsilon^2)\left(1 - \frac{\epsilon}{2\eta}\right)\eta\right). \quad (12.9)$$

When $\epsilon = 1$, the last expression reduces to (12.7) as could be expected. In such a case, $P(x,y,t)$ is a multivariate Gaussian function, whose width scales as $t^{3/2}$ in the x direction, while having a normal diffusive behavior in the y direction.

A numerical analysis of the integrand in (12.8) indicates that a reasonable approximation is to neglect the 2nd term in the prefactor square root, which is completely justified for small values of t. When t is large, it seems more adequate to neglect the first term in the square root, leading to an exact solution in terms of the parabolic cylinder functions $D_{1/2}(z)$ (Abramowitz and Stegun (1964)). However, due to the compensation introduced by the slow damped integrand's oscillations, the asymptotic behavior of this solution is coincident with the result of neglecting the 2nd term in the prefactor square root (Wio and Zanette (1993)). Within this approximation (12.8) reduces to

$$P(x,y,t) \approx \frac{\eta^2}{v_o dt^2} \frac{1}{\sqrt{4\pi da^2 t}} \left(\frac{4dt}{\phi(\eta,\epsilon)}\right)^{1/2} \exp\left\{-\frac{\eta^2}{4dt}\right\}$$
$$\times \exp\left\{-\left(\frac{\eta^2}{dv_o t^2}\right)^2 \frac{dt}{\epsilon\phi(\eta,\epsilon)}\psi(x,\eta,\epsilon)\right\}. \quad (12.10)$$

Finally, and in order to correct for the jump at $y = \epsilon a$ due to the approximation indicated above, we include in the last expression the factor $P(x,\epsilon a^-\epsilon,t)/P(x,\epsilon a^+\epsilon,t)$ and normalize the whole distribution to one. Figure 12.1 you can see a sketch of the marginal conditional probability $P(x,t) = \int dy P(x,y,t)$ for a fixed time and different values of ϵ, while Fig. 12.2 shows a sketch, for a fixed value of ϵ and three different times.

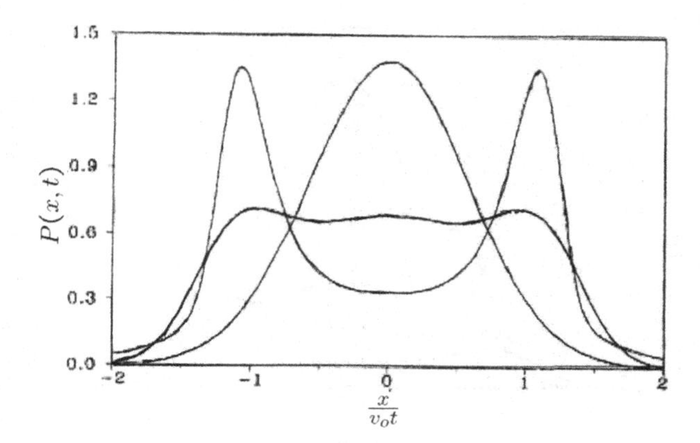

Fig. 12.1 Sketch of the marginal conditional probability $P(x,t)$ as a function of the variable $x/v_o t$, at $t = 0.5$, and for three different values of ϵ: from top to bottom $\epsilon = 1., 0.6, 0.3$. Other parameters are $a = d = v_o = 1$.

We will not extent this discussion further, but refer the reader to Wio and Zanette (1993), where these results were numerically evaluated. There, it was also shown that in addition to the above indicated mono- to bistable behavior as β is varied a transient tri-modality was also found (see for instance, Fig. 12.2), that was numerically verified by means of an elaborate numerical analysis of (12.1). Also, the critical value of β results to be $\beta \approx 0.75$, in good agreement with the previously indicated numerical value.

12.2 Diffusion Controlled Reactions

The model we want to discuss in this section was originally used in neutron diffusion problems and exploited to discuss diffusion controlled reactions (Abramson *et al.* (1991b)). In the following we present the model, a standard method of solution using stochastic techniques as well as a path-integral approach (Wio and Nicolini (1993b); Sánchez *et al.* (2000)). The latter offers the possibility of extending the results of this model from the case $A + B \rightarrow B$ to the $A + B \rightarrow 0$ one.

12.2.1 *The Model*

For sake of completeness, we briefly review here a few of the results of (Rodriguez *et al.* (1993); Abramson *et al.* (1991b)). We consider the case of a one-dimensional diffusion equation for the processes $A + B \rightarrow B$ modeled

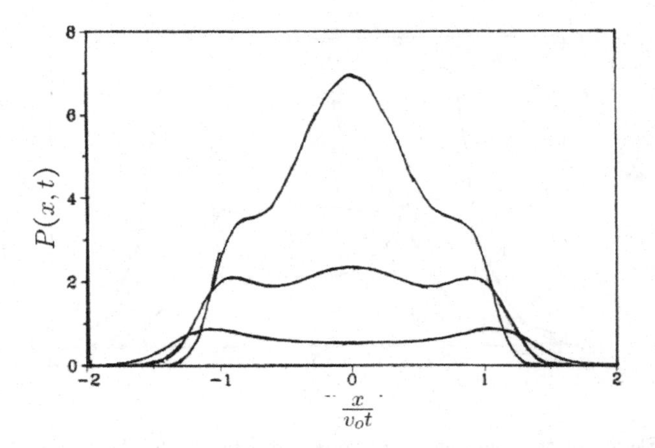

Fig. 12.2 Sketch of the marginal conditional probability $P(x,t)$ as a function of the variable $x/v_o t$, at a fixed value of $\epsilon = 0.5$ and three times: from top to bottom $t = 0.1, 0.2, 0.5$. Other parameters are $a = d = v_o = 1$. The tri-modality is apparent.

according to this theory

$$\partial_t N(x,t) = D\partial_x^2 N(x,t) - \gamma\delta(x - \epsilon(t))N(x,t) + S_0(x,t), \qquad (12.11)$$

where $N(x,t)$ is the density of the A reactant, γ is a reaction constant, $\epsilon(t)$ is the random position of the B reactant (that here plays the role of the absorber), and $S_0(x,t)$ is an independent source of A particles that here we will consider as zero. The only hypothesis that is necessary to take into account is that the processes $\epsilon(t)$ is Markovian. As we are interested in calculating $\langle N(x,t)\rangle$, that is the averaged particle density, calling $n(x,t) = \langle N(x,t)\rangle$ and taking averages in (12.11), we obtain

$$\partial_t n(x,t) = D\partial_x^2 n(x,t) - \gamma A(x,t) + S_0(x,t), \qquad (12.12)$$

where $A(x,t) = \langle\delta(x - \epsilon(t))N(x,t)\rangle$. The effect of the reactant B on A is completely specified by $A(x,t)$. If this quantity is known it is simple to evaluate $n(x,t)$. Taking the integral form of (12.11)

$$\begin{aligned} N(x,t) = &\int dx' K^{(0)}(x,t|x',0)N(x',0) \\ &+\int_0^t dt' \int dx' K^{(0)}(x,t|x',t')S_0(x',t') \\ &-\gamma\int_0^t dt' \int dx' K^{(0)}(x,t|x',t')\delta(x' - \epsilon(t'))N(x',t'). \end{aligned} \qquad (12.13)$$

In the following we adopt

$$S_o(x,t) = N(x,0)\delta(t) . \tag{12.14}$$

Iterating now (12.13), multiplying by $\delta(x-\epsilon(t))$ and taking averages, we find:

$$\begin{aligned} A(x,t) &= \langle \delta(x-\epsilon(t))N(x,t)\rangle \\ &= \int_0^t dt' \int dx' K^{(0)}(x,t|x',t')\langle \delta(x-\epsilon(t))\rangle S(x',t') \\ &\quad -\gamma \int_0^t dt' \int dx' \int_0^{t'} dt'' \int dx'' \langle \delta(x-\epsilon(t))\delta(x'-\epsilon(t'))\rangle \\ &\quad \times K^{(0)}(x,t|x',t')K^{(0)}(x',t'|x'',t'')S(x'',t'') + \ldots . \end{aligned} \tag{12.15}$$

As ϵ is a Markov process we have

$$\begin{aligned} \langle \delta(x-\epsilon(t))\rangle &= \int \delta(x-\epsilon)P(\epsilon,t)d\epsilon \\ &= P(x,t) \\ &= \int \omega(x,t|x_0,0)P(x_0)dx_0 , \end{aligned} \tag{12.16}$$

and

$$\begin{aligned} \langle \delta(x-\epsilon(t))\delta(x'-\epsilon(t'))\rangle &= P(x,t;x',t') \\ &= \omega(x,t|x',t')\int \omega(x,t|x_0,0)P(x_0)dx_0 , \end{aligned} \tag{12.17}$$

where we have used the standard notation for the joint and the initial probability distributions, and $\omega(x,t|x',t')$ is the conditional probability for the process ϵ. Substitution of these results into (12.15) leads to a closed equation for $A(x,t)$:

$$\begin{aligned} A(x,t) &= \int_0^t dt' \int dx' K^{(0)}(x,t|x',t')S(x',t') \int \omega(x,t|x_0,0)P(x_0)dx_0 \\ &\quad -\gamma \int_0^t dt' \int dx' K^{(0)}(x,t|x',t')\omega(x,t|x',t')A(x',t') . \end{aligned} \tag{12.18}$$

With this result, we are in principle able to solve (12.12) for $n(x,t)$.

12.2.2 *Point of View of Path Integrals*

The problem posed by (12.11) is equivalent to a Schrödinger equation (in imaginary time) with a δ potential. The path-integral representation of its solution could be written as (Wio and Nicolini (1993b))

$$N(x_f, t_f) = \int dx_0 K(x_f, t_f|x_0, 0) N(x_0, 0), \tag{12.19}$$

where

$$K(x_f, t_f|x_0, 0) = \int \mathcal{D}[x]\, e^{-S[x]}, \tag{12.20}$$

with $\mathcal{D}[x]$ the usual measure. The action $S[x]$ has the form

$$S[x] = S^{(0)}[x] + S^{(1)}[x] \tag{12.21}$$

$$S^{(0)}[x] = \frac{1}{4D}\int_0^{t_f} \dot{x}^2(t)\, , dt \tag{12.22}$$

$$S^{(1)}[x] = \int_0^t dt \gamma \delta(x(t) - \epsilon(t)) \tag{12.23}$$

Expanding the exponent in (12.20) in powers of γ

$$e^{-S[x]} = e^{-S^{(0)}[x]} \left\{ 1 + \sum_{n=1}^{\infty} \frac{(-1)^n}{n!} (S^{(1)}[x])^n \right\} . \tag{12.24}$$

(12.20) can be written as

$$K(x_f, t_f|x_0, 0) = K^{(0)}(x_f, t_f|x_0, 0) + K^{(1)}(x_f, t_f|x_0, 0) \, , \tag{12.25}$$

$$K^{(0)}(x_f, t_f|x_0, 0) = \int \mathcal{D}[x]\, e^{-S^{(0)}[x]} \, , \tag{12.26}$$

$$K^{(1)}(x_f, t_f|x_0, 0) = \sum_{n=1}^{\infty} (-\gamma)^n G^{(n)}(x_f, t_f|x_0, 0) \, . \tag{12.27}$$

where

$$G^{(n)}(x_f, t_f|x_0, 0) = \frac{1}{n!}\int \mathcal{D}[x]\, (S^{(1)}[x])^n e^{-S^{(0)}[x]}, \tag{12.28}$$

with

$$(S^{(1)}[x])^n = \int_0^{tf} dt_n \int_0^{tf} dt_{n-1} \dots \int_0^{tf} dt_1 \prod_{k=1}^{n} \delta(x(t_k) - \epsilon(t_k)) \, . \tag{12.29}$$

As is very well known, ordering in time ($t_1 < t_2 < \cdots < t_n$), and changing integration limits, we may rewrite (12.29) as

$$(S^{(1)}[x])^n = n! \int_0^{tf} dt_n \int_0^{tn} dt_{n-1} \cdots \int_0^{t2} dt_1 \prod_{k=1}^{n} \delta(x(t_k) - \epsilon(t_k)) . \tag{12.30}$$

Now, with the help of the δ's, we make a partition of the time interval in $G^{(n)}$, which gives

$$G^{(n)}(x_f, t_f|x_0, 0) = \int_0^{tf} dt_n \ldots \int_0^{t3} dt_2 \int_0^{t2} dt_1 K^{(0)}(x_f, t_f|\epsilon_n, t_n) \times \left[\prod_{k=1}^{n} K^{(0)}(\epsilon_k, t_k|\epsilon_{k-1}, t_{k-1}) \right] K^{(0)}(\epsilon_1, t_1|x_0, 0) . \tag{12.31}$$

At this point we must resort to an average over the processes ϵ. Such averaging, following ideas similar to the ones used previously, lead us to

$$n(x,t) = \int dx_0 \langle K(x_f, t_f|x_0, 0)\rangle N(x_0, 0), \tag{12.32}$$

with

$$\langle K(x_f, t_f|x_0, 0)\rangle = \langle K^{(0)}(x_f, t_f|x_0, 0)\rangle + \langle K^{(1)}(x_f, t_f|x_0, 0)\rangle , \tag{12.33}$$

$$\langle K^{(0)}\rangle = K^{(0)}(x_f, t_f; x_0, 0) , \tag{12.34}$$

$$\langle K^{(1)}\rangle = \sum_{n=1}^{\infty} (\gamma)^n \langle G^{(n)}(x_f, t_f|x_0, 0)\rangle . \tag{12.35}$$

Considering (12.31) and (12.35) we find

$$\langle G^{(n)}(x_f, t_f|x_0, 0)\rangle = \int d\epsilon_n \ldots \int d\epsilon_1 \int_0^{t_f} dt_n \ldots \int_0^{t_2} dt_1 \times K^{(0)}(x_f, t_f|\epsilon_n, t_n) \left[\prod_{k=1}^{n} K^{(0)}(\epsilon_k, t_k|\epsilon_{k-1}, t_{k-1}) \right] \times K^{(0)}(\epsilon_1, t_1|x_0, 0) P(x_0, 0; \epsilon_1, t_1 \ldots \epsilon_n, t_n, x_f, t_f) . \tag{12.36}$$

Now, from (12.17), for the joint probability distribution we have

$$P(x_0, 0; \epsilon_1, t_1 \ldots \epsilon_n, t_n, x_f, t_f)$$
$$= \omega(x_f, t_f|\epsilon_n, t_n) \left[\prod_{k=1}^{n} \omega(\epsilon_k, t_k|\epsilon_{k-1}, t_{k-1})\right] \omega(\epsilon_1, t_1|x_0, 0)\,. \quad (12.37)$$

If we also integrate over the initial distribution of reactant A, i.e. $N(x_0, 0)$, we obtain

$$\langle\langle G^{(n)}(x_f, t_f|x_0, 0)\rangle\rangle = \int dx_0 \langle G^{(n)}(x_f, t_f|x_0, 0) N(x_0, 0)\rangle$$
$$= \int d\epsilon_n \ldots \int d\epsilon_1 \int_0^{t_f} dt_n \ldots \int_0^{t_2} dt_1$$
$$\times K^{(0)}(x_f, t_f|\epsilon_n, t_n)\omega(x_f, t_f|\epsilon_n, t_n)$$
$$\times \left[\prod_{k=2}^{n} K^{(0)}(\epsilon_k, t_k|\epsilon_{k-1}, t_{k-1})\, \omega(\epsilon_k, t_k|\epsilon_{k-1}, t_{k-1})\right]$$
$$\times \int dx_0 \int dy_0 K^{(0)}(\epsilon_1, t_1|x_0, 0)\omega(\epsilon_1, t_1|y_0, 0)$$
$$\times N(x_0, 0) P(y_0)\,. \quad (12.38)$$

This result is completely coincident (making a term by term comparison) with the one we would obtain by solving (12.12) via a path-integral method. However, there are more direct ways of obtaining such a path-integral solution, but the above procedure seems to be the most adequate one in order to allow us to consider the case $A + B \to 0$. In such a case, we need to solve a coupled set of equations of the form (12.11), implying that $\omega(x, t|x', t')$ will not be fixed externally but must be obtained in a selfconsistent way solving this system (i.e. it would not necessarily result to be Gaussian).

12.2.3 *Results for the Reaction $A + B \to B$*

As we said before, we consider the reaction example $A + B \to B$, where A plays the role of the absorbed element and B of the absorbent. Both A and B have a diffusive behavior. We will consider the following situations : in the first case B is concentrated in the origin and A is uniformly distributed; in the second case, both A and B are uniformly distributed. We have for the Green's function and for the conditional probability

$$K^{(0)}(x, t|x', t') = \frac{1}{\sqrt{4\pi D(t - t')}} \exp\left[-\frac{(x - x')^2}{4D(t - t')}\right], \quad (12.39)$$

and

$$\omega(x,t|x',t') = \frac{1}{\sqrt{4\pi d\,(t-t')}} \exp\left[-\frac{(x-x')^2}{4d(t-t')}\right] . \tag{12.40}$$

The initial condition for A is $S(x',t') = N_0\delta(t')$ for both cases. For B we have

$$P_1(x_0,t_0) = P_0\delta(x_0) , \qquad P_2(x_0,t_0) = P_0 . \tag{12.41}$$

Replacing this in (12.18), and transforming Fourier–Laplace in space and time respectively, we obtain

$$A_1(k,p) = \frac{N_0 P_0}{\sqrt{2\pi}(p+k^2 d)} \frac{\sqrt{p+k^2\beta}}{\sqrt{p+k^2\beta}+\gamma/\sqrt{8\pi(D+d)}} \tag{12.42}$$

and

$$A_2(k,p) = \frac{N_0 P_0 \delta(k)\sqrt{2\pi}}{p} \frac{\sqrt{p+k^2\beta}}{\sqrt{p+k^2\beta}+\gamma/\sqrt{8\pi(D+d)}} \tag{12.43}$$

where $\beta = \frac{Dd}{(D+d)}$.

The integral form of the (12.12) for the density n is

$$n = N_0 - \gamma \int_0^t dt' L^{-1}[A(0,p),p,t'] , \tag{12.44}$$

where L^{-1} stands for the inverse Laplace transform. In the last equation we took $k=0$ in order to obtain the total density. As we want to obtain the relative density we adopt $\Delta n = n/n_0$ where n_0 is the value that we obtain from n making $\gamma = 0$ (i.e., without absorption). Then, we have

$$\Delta n = 1 - \sqrt{32\pi}\alpha P_0 (D+d)\gamma^{-1}\left(2\sqrt{\tau/\pi} + e^{\tau}\mathrm{Erfc}\,(\sqrt{\tau}) - 1\right) , \tag{12.45}$$

where $\tau = \gamma^2 t/[8\pi(D+d)]$. In the first case $\alpha = 1$ while in the second $\alpha = 2\pi$. It is worth to stress the fact that these solutions do not have the usual exponential or inverse power law decreasing behavior, instead they vanish at a finite time.

Finally let us consider the following case. Assume the reactants B do not diffuse, but have a fixed spatial distribution according to

$$P_B(x) = N_B(1 + \cos(k_0 x)).$$

The solution of the problem, where now we must take $\omega(x,t|x',t') = \delta(x - x')$, have the characteristic that, $n(x,t)$, after a short initial transient, reaches a form (decaying in time) with a spatial modulation of wave-number k_0, that corresponds to the *conjugate* of $P_B(x)$. Such result is the signature of segregation (originated by fluctuations), and when treating, within a similar approach, the case $A + B \to 0$, we have found (in a low order approximation) a similar result.

Chapter 13

What was Left Out

. . . .
not because the calculation of the apostle is wrong, but because we have not learnt the art on which such calculus is grounded.
. . . .

Umberto Eco

In these notes we have presented some elements of path integration techniques as applied to stochastic processes. We have focused on the description of Markov processes, and have also taken a glimpse of non-Markov processes. However, due to space limitations, there are several interesting and related topics that have been left out. In order to help the reader and/or students, and for the sake of completeness, we will briefly mention some of them and also indicate some relevant bibliography.

• Statistical Physics: there are direct applications that makes it possible to obtain the density operator or the partition function (Feynman (1972); Schulman (1981); Wiegel (1986)). One very interesting and related technique is the Hubbard-Stratonovich transformation. It is a functional technique that allows to rewrite the partition function in a representation where, with the assistance of some auxiliary fields, the nonlinear terms can be linearized. Some examples of application are in connection with magnetic impurities (Evenson *et al.* (1970); Morandi, *et al.* (1974); Wio *et al.* (1978)); and semiconductor-metal transitions and alloy systems (Wio (1983)).

• Polymers: there have been numerous studies making the connection between path integrals and different kinds of random-walk (Wiegel (1986); Kleinert (1990b)).

• Trotter formula: as we have seen, this is an "easy" way to derive a *phase-space representations* (in contrast to the *configuration space* representation, see Marinov (1980)) of the path integral (Schulman (1981); Wio (1988)). It was also largely exploited as the starting point of numerical procedures.

• Representations for discrete variables: some examples of which are spins, composed stochastic processes (van Kampen (2004)) including continuous and discrete variables. A very complete presentation can be found in Langouche *et al.* (1982).

• Problems with Fermionic-like degrees of freedom: this involves the treatment of Grassman variables and is also related to coherent state representations (Schulman (1981); Das (1994)).

• Brownian motion. In addition to the cases discussed in these notes, there are an extremely large number of situations with several applications that we will not numerate here, but only indicate the work by Chatterjee and Cherayil (2010).

• Numerical methods: there is an extensive bibliography on numerical methods in general and for propagation in particular. We indicate here a few, due to the didactic presentation or because it is related work by the author: Salem and Wio (1986); MacKeown (1985); Gerry and Kiefer (1988); Abramson *et al.* (1991a); Hassan *et al.* (1994). It is worth to here indicate the review-like work of Bernu and Ceperley (2002).

• We have neither commented on the effect of the noise spectrum as, for instance, discussed in (Dykman and Lindenberg (1994)), nor the discussed the distribution of fluctuational paths as discussed in (Dykman and Smelyanskiy (1998)).

• Field theory approaches: this is an area where path integration has been extremely fruitful, from both quantal and classical points of view. Some didactic presentations are Felsager (1985); Das (1994). We emphasize the applications in relation with reaction-diffusion and dynamical systems (Graham (1978); Freidlin (1985); Graham (1987); Foster and Mikhailov (1988); Fedotov (1993); Izús *et al.* (1995, 1996, 1998)), and the utilization of these results to describe stochastic resonance in extended systems (Wio (1996); Wio and Castelpoggi (1997); Castelpoggi and Wio (1997, 1998); Kuperman *et al.* (1998); Wio *et al.* (1998)). It is worth to here stress the formalism discussed in Martin, *et al.* (1973); Janssen (1976); DeDominicis and Peliti (1978).

• Perturbation techniques has been at the heart of the Feynman approach, and almost everything that has been done in quantum mechanics or field theory could be done here (Langouche *et al.* (1982); Schulman (1981); Dickman and Vigidal (2003)).

• Somehow related with the perturbation techniques indicated above we can highlight the so called Ω expansion technique due to van Kampen (van Kampen (2004)), that was discussed through a path integral approach in (Calisto and Tirapegui (1993)).

• Another very interesting technique is the one associated to *instantons* (Coleman (1977); Felsager (1985)). Such technique, that is so useful in quantum mechanics and field theory, have an equivalent within a stochastic framework to obtain mean passage times.

• A problem very much related with our discussion on fractional process is the study of the *fractional Schrödinger equation* (Garbaczewski (2010); Lenzi *et al.* (2010)). Also related with *fractional motions* is the work on the path integral approach to the *Continuous Time Random Walk* (Eule and Friedrich (2011, 2012)).

• We have not analyzed the effect of boundary conditions. In the quantum case, this aspect was analyzed in detail in Kleinert (1990b), while in Hassan *et al.* (1994) a propagation problem with Dirichlet, Neumann and albedo boundary conditions was discussed.

The list could continue further, but we believe that the material included in these notes as well as the subjects indicated above, can already offer a feeling and attract attention to the many interesting facets of this technique. We hope that these notes, and the course on which they are based, will awaken the curiosity of the students and stimulate them to delve deeper into the diverse aspects connected with path integrals and perhaps apply such an approach in their own research.

I left for the various futures (but not all of them) my garden of the bifurcating paths......
Jorge Luis Borges

Appendix A

Space-Time Transformation: Definitions and Solutions

A.1 Definitions

The quantities a_N, p_N and q_N are defined according to

$$a_N = \beta \prod_{j=1}^{N-1} \frac{\beta}{2\gamma_j}; \quad \gamma_1 = \alpha_1, \quad \gamma_j = \alpha_j - \frac{\beta^2}{4\gamma_{j-1}}$$

$$p_N = -\frac{\beta}{2} + \sum_{j=1}^{N-1} \frac{\beta_j^2}{4\gamma_j}; \quad q_N = -\frac{\beta}{2} + \frac{\beta^2}{4\gamma_{N-1}};$$

$$\beta_1 = \beta; \quad \beta_j = \beta \prod_{k=1}^{j-1} \frac{\beta}{2\gamma_k}, \tag{A.1}$$

In order to determine the limiting (when $N \to \infty$) values of a_N, p_N, and q_N, it is useful to define the following auxiliary quantities

$$\lambda_j = \frac{2}{\beta}\gamma_j \; ; \; \Lambda_k = \prod_{j=1}^{k} \frac{1}{\lambda_j}, \tag{A.2}$$

with α_j as defined after (6.14) and γ_j adopting the form

$$\gamma_j = \beta\left(1 + \frac{\varepsilon^2 w_j}{2}\right) - \frac{\beta^2}{4\gamma_{j-1}}, \tag{A.3}$$

that allows us to obtain the following equation for λ_j

$$\lambda_j = 2\left(1 + \frac{\varepsilon^2 w_j}{2}\right) - \frac{1}{\lambda_{j-1}}. \tag{A.4}$$

If we now define that $\lambda_j = \frac{Q_{j+1}}{Q_j}$, the last equation can be rewritten as

$$Q_{j+1} - 2Q_j + Q_{j-1} = \omega_j \varepsilon^2 Q_j, \tag{A.5}$$

which, in the limit $N \to \infty$ (and $\varepsilon \to 0$), becomes (6.16), with the initial condition $Q_0 = Q(t_a) = 0$, that follows from (A.5).

Finally, we can express the coefficients a_N, p_N, and q_N as functions of the new variables

$$a_N = \beta\Lambda_{N-1} = \beta\frac{Q_1}{Q_N} = \beta\frac{Q_1 - Q_0}{Q_N} = \beta\varepsilon\left(\frac{Q_1 - Q_0}{\varepsilon}\right)\frac{1}{Q_N}, \tag{A.6}$$

$$p_N = -\frac{\beta}{2}\left(1 - \sum_{j=1}^{N-1}\frac{Q_1^2}{Q_{j+1}Q_j}\right), \tag{A.7}$$

$$q_N = -\frac{\beta\varepsilon}{2}\left(\frac{Q_N - Q_{N-1}}{\varepsilon Q_N}\right). \tag{A.8}$$

A.2 Solutions

The replacement of the general solution for $Q(t)$ indicated in (6.22) into (6.17,6.18,6.19), leads us to obtain the limiting values of a_N, p_N and q_N. For p_N we find

$$\lim_{N\to\infty} p_N = \lim_{\varepsilon\to 0}\frac{-1}{2D}\left(\frac{1}{\varepsilon} + \frac{1}{\int_{t_a}^{t_b} d\tau\, e^{2\int_{t_a}^{\tau} a(\varsigma)d\varsigma}} - \frac{1}{\int_{t_a}^{t_a+\varepsilon} d\tau\, e^{2\int_{t_a}^{\tau} a(\varsigma)d\varsigma}}\right). \tag{A.9}$$

Making a Taylor expansion up to second order in ε of the last denominator we can calculate the limit in (A.9) yielding

$$\lim_{N\to\infty} p_N = \frac{-1}{2D}\left(\frac{1}{\int_{t_a}^{t_b} d\tau\, e^{2\int_{t_a}^{\tau} a(\varsigma)d\varsigma}} + a(t_a)\right). \tag{A.10}$$

The expressions for a_N and q_N, in terms of the explicit form for $Q(t)$ results in

$$\lim_{N\to\infty} q_N = \frac{-1}{2D}\left(\frac{e^{\int_{t_a}^{t_b} a(\varsigma)d\varsigma}}{\int_{t_a}^{t_b} d\tau\, e^{2\int_{t_a}^{\tau} a(\varsigma)d\varsigma}} - a(t_b)\right) \tag{A.11}$$

$$\lim_{N\to\infty} a_N = \frac{1}{D}\frac{e^{\int_{t_a}^{t_b} a(\varsigma)d\varsigma}}{\int_{t_a}^{t_b} d\tau\, e^{2\int_{t_a}^{\tau} a(\varsigma)d\varsigma}}. \tag{A.12}$$

Appendix B

Basics Definitions in Fractional Calculus

References [Oldham and Spanier (1974); Podlubny (1998)] are among the most appropriate texts to study fractional calculus, from where we have taken the following definitions.

The fractional derivative of order $\alpha > 0$, that is denoted by $\frac{d^\alpha}{dt^\alpha} \equiv {}_aD_t^\alpha$, can be defined in terms of its inverse, the fractional integral, indicated by ${}_aD_t^{-\alpha}$

$$ {}_aD_t^{-\alpha}h(t) = \frac{1}{\Gamma(\alpha)} \int_a^t [t - t']^{\alpha-1} h(t')dt'. \tag{B.1}$$

The fractional derivative of arbitrary order ${}_aD_t^\alpha$, where $n - 1 \leq \alpha < n$, could be defined through the fractional integration of order $n - \alpha$, and successive ordinary derivatives of order n

$$ {}_aD_t^\alpha h(t) = \left(\frac{d}{dt}\right)^n {}_aD_t^{\alpha-n}h(t). \tag{B.2}$$

When $0 < \alpha < 1$ or $N = 1$, one obtains

$$ {}_aD_t^\alpha h(t) = \frac{1}{\Gamma(\alpha)} \left(\frac{d}{dt}\right) \int_a^t [t - s]^{\alpha-1} h(s)ds. \tag{B.3}$$

The above indicated definitions are known as *Riemann–Liouville* (RL) or *Weyl* integral and derivative if $a = 0$ and $a = -\infty$, respectively.

Summarizing, let us assume that $g(x)$ is a real ($g : \mathbb{R} \to \mathbb{R}$) and *well behaved function*. As indicated above, the *Riemann–Liouville fractional integral operators* of order α are defined by

$$ {}_aD_t^{-\alpha}g(t) := \frac{1}{\Gamma(\alpha)} \int_a^t [t - s]^{\alpha-1} g(s)ds, \tag{B.4}$$

$$ {}^bD_t^{-\alpha}g(t) := \frac{1}{\Gamma(\alpha)} \int_t^b [t - s]^{\alpha-1} g(s)ds, \tag{B.5}$$

while *Riemann–Liouville fractional differential operators* of order α are defined by

$$
{}_aD_t^\alpha g(t) := \frac{(-1)^m}{\Gamma(m-\alpha)} \frac{d^m}{dt^m} \int_a^t [t-s]^{-\alpha+m-1} g(s) ds, \tag{B.6}
$$

$$
{}^bD_t^{-\alpha} g(t) := \frac{1}{\Gamma(\alpha)} \int_t^b [t-s]^{-\alpha+m-1} g(s) ds, \tag{B.7}
$$

where m is an integer number such that $m-1 \leq \alpha < m$.

For completeness, the *Riesz fractional differential operator* is defined by the symmetric combination

$$
\frac{\partial^\alpha}{\partial |s|^\alpha} := \frac{-1}{2\cos(\frac{\pi\alpha}{2})} \left[{}_{-\infty}D_s^\alpha + {}^{\infty} D_s^\alpha \right]. \tag{B.8}
$$

Bibliography

Abramowitz M., Stegun I.A. (1964). *Handbook of Mathematical Functions* (Dover, New York) ISBN.0-486-61272-4.

Abramson G., Wio H.S., Salem L.D, (1991). in *Nonlinear Phenomena in Fluids, Solids and other Complex Systems*, Cordero P., Nachtergaele B., Eds., (North-Holland, Amsterdam).

Abramson G., Bru Espino A., Rodriguez M.A.,Wio H.S., (1994). Phys. Rev. E. **50**, 4319.

Ajanapon P., (1987). Am. J. Phys. **55**, 159.

Batista C.D., Drazer G., Reidel D., Wio H.S., (1996). Phys. Rev. E **54**, 86.

Bernu B., Ceperley D.M., (2002). in *Quantum Simulations of Complex Many Body Systems. From Theory to Algorithms*, Grotendorst J., Marx D., Muramatsu A., Eds. J.von Neumann Institute for Computing, NIC Series **10**, 51-61.

Bezrukov S.M., Vodyanoy I. (1995). Nature **378**, 362.

Borland L. (1998). Phys. Lett. A **245**, 67.

Borland L.(1998). Phys. Rev. E **57**, 6634.

Brink D.M., (1985). *Semiclassical Methods for Nucleus-Nucleus Scattering*, (Cambridge U.P., Cambridge).

Buchdahl H.A., (1988). J. Math. Phys. **29**, 1122.

Burkitt A.N. (2006). Biological Cybernetics **95**, 97-112.

Calisto H., and Tirapegui E., (1993). J. Stat. Phys. **71**, 683.

Calisto H., Mora F. and Tirapegui E., (2006). Phys. Rev. E **74**, 022102.

Calvo I. and Sánchez R. (2008), J. Phys. A: Math. Theor. **41**, 282002.

Calvo I., Sánchez R. and Carreras B.A., J. Phys. A: Math. Theor. **42**, 055003.

Castelpoggi F., Wio H.S., (1997). Europhysics Lett. **38**, 91.

Castelpoggi F. and Wio H. S. (1998). Phys. Rev. E **57**, 5112.

Castro F., Wio H.S., Abramson G., (1995). Phys. Rev. E **52**, 159.

Castro F., Sanchez A., Wio H.S., (1995). Phys. Rev. Lett. **75**, 1691.

Castro F.J., Kuperman M.N., Fuentes M.A., Wio H.S. (2001). Phys. Rev. E **64**, 051105.

Cerdeira H.A., Lundqvist S., Mugnai D., Ranfagni A., Sa-yakanit V., Schulman L.S., Eds, (1992). *Path Integration. Trieste 1991*, (World Scientific, Singapore).

Chatterjee D. and Cherayil B. (2010), Phys. Rev. E **82**, 051104.
Coleman S. (1977). *The Uses of Instantons*, in Proc. IV Int. School of Subnuclear Physics, Erice, Italy, Ed. A. Zichici, (Academic Press, N.Y.).
Colet P., Wio H. S. and San Miguel M. (1989). Phys Rev **A39**, 6094.
Cruz M.G., (1992). Am. J. Phys. **60**, 1127.
Curado E.M.F., Tsallis C. (1991). J. Phys. A **24**, L69.
Curado E.M.F., Tsallis C. (1991). J. Phys. A **24**, 3187.
Curado E.M.F., Tsallis C. (1992). J. Phys. A **25**, 1019.
Chernyak V.Y., Chertkov M., Jarzynski C., (2006). J. Stat. Mech. P08001.
Das A., (1994). *Field theory. a path integral approach* (World Scientific, Singapore).
Davison B., (1954). Proc. Roy. Soc. A **225**, 252.
DeDominicis C., Peliti L., (1978). Phys. Rev. B **18**, 353-376.
Deininghaus U., Graham R., (1979). Z. Phys. B **34**, 211.
Dickman R, Vigidal, R. (2003). Braz. J. Phys. **33**, 73.
Donoghue J.F., Holstein B.R., (1988). Am.J.Phys. **56**, 216.
Duru I.H., Kleinert H., (1979). Phys. Lett. B **84**, 30.
Duru I.H., Kleinert H., (1982), Fortschr. Physik **30**, 401.
Duru I.H., (1983). Phys. Rev. D **28**, 2689.
Dykman M. and Lindenberg K., (1994). *Fluctuations in Nonlinear Systems Driven by Colored Noise*, in *Some Problems in Statistical Physics*, Ed.G. Weiss (SIAM, Philadelphia).
Dykman M. and Smelyanskiy V.N. (1998), SuperLatt. and Struct. **23**, 495.
Eab Ch.H. and Lim S.C., (2006) Physica A **371**, 303-316.
Eliazar I. I. and Shlesinger M. F., (2012), J. Phys. A: Math. Theor. **45** 162002.
Eliazar I. I. and Shlesinger M. F., (2012), *Fractional Motions*, in press.
Eule S. and Friedrich R., (2011), *Path integral formulation of anomalous diffusion processes*, arXiv:1110.5771v1.
Eule S. and Friedrich R., (2012), *Path integral formulation and path probabilities of CTRW, 25th M. Smoluchowski Symposium on Statistical Physics*, Krakow, Poland, Sept.2012.
Evans D.J., Searles D.J. (2002). Adv. in Phys. **51**, 1529-1585.
Evenson W.E., Schrieffer J.R., Wang S.Q., (1970). J. Appl. Phys. **41**, 1199.
Fedotov S.P. (1993). Phys. Lett. A **176**, 220.
Felsager B., (1985). *Geometry, Particles and Fields* (Odense University Press, Odense).
Feynman R.P., (1948). Rev. Mod. Phys. **20**, 367.
Feynman R.P. and Vernon F.L., (1963). Ann. Phys. (N.Y.) **24**, 118.
Feynman R.P., Hibbs A.R., (1965). *Quantum mechanics and path integrals* (McGraw-Hill, New York).
Feynman R.P., (1972). *Statistical Mechanics* (Benjamin, Massachussetts).
Fletcher G., (1990). Am. J. Phys. **58**, 50.
Foster A., Mikhailov A.S. (1988). Phys. Lett. A **126**, 459.
Fox R. F. (1986), Phys. Rev. A **33**, 467.
Freidlin M. (1985). *Functional Integration and Partial Differential Equations*, (Princeton University Press, Princeton).

Fuentes M.A., Toral R., Wio H.S. (2001a). Physica A **295**, 114.
Fuentes M.A., Toral R., Wio H.S. (2001b). Physica A **303**, 91.
Gammaitoni L., Hänggi P., Jung P. and Marchesoni F., (1998). Rev. Mod. Phys. **70**, 223.
Garbaczewski P. (2010), Central Europ. J. Phys. **8**, 699-708.
Gardiner C.W. (2009). *Handbook of Stochastic Methods, 4th Ed.* (Springer-Verlag, Berlin).
Garrison J.C., Wright E.M. (1985). Phys. Lett. A **108**, 129.
Gell-Mann M., Tsallis C. (Eds.), (2003). *Nonextensive Entropy-Interdisciplinary Applications* (Oxford U.P., Oxford).
Gerry Ch.C., Kiefer J., (1988). Am. J. Phys. **56**, 1002.
Goldstein H. (1980). *Classical Mechanics*, (Addison Wesley, Reading Mass.).
Graham R. (1978). in *Lecture Notes in Physics*, vol. 84 (Springer–Verlag, Berlin).
Graham R. (1987). in *Instabilities and Nonequilibrium Structures*, Eds.E.Tirapegui and D.Villaroel (D.Reidel, Dordrecht).
Grosche C., Steiner F. (1998). *Handbook of Feynman Path Integrals* (Springer).
Grosjean C.C., (1988). J.Comp.and Appl.Math. **21**, 311 and **23**, 199.
Haken H. (1978). *Synergetics. An Introduction*, 2nd Ed. (Springer-Verlag, N.Y.).
Hänggi P., Thomas H., (1975). Z.Phys. B **22**, 295.
Hängi P., (1989). Z. Phys. B **75**, 275.
Hängi P., Jung P. and Marchesoni F. (1989). J. Stat. Phys. **54**, 1367.
Hängi P., Talkner P., Borkovec M. (1990). Rev. Mod. Phys, **62**, 251.
Hängi P., and Jüng, P. (1995), Advances in Chemical Physics **89**, 239326.
Hassan S.A., Kuperman M.N., Wio H.S., Zanette D.H. (1994). Physica A **206**, 380.
Ho R., Inomata A., (1982). Phys. Rev. Lett. **48**, 231.
Holstein B.R., (1981a). Am.J.Phys. **51**, 897.
Holstein B.R., (1981b). Am.J.Phys. **51**, 1015.
Holstein B.R., (1982). Am.J.Phys. **52**, 321.
Holstein B.R., (1983). Am.J.Phys. **53**, 723.
Holstein B.R., (1988). Am.J.Phys. **56**, 338.
Holstein B.R., (1988). Am.J.Phys. **56**, 894.
Holstein B.R., (1989). Am.J.Phys. **57**, 714.
Holstein B.R., Swift A.R., (1982a). Am.J.Phys. **50**, 829.
Holstein B.R., Swift A.R., (1982b). Am.J.Phys. **50**, 833.
Horsthemke W., Lefever R. (1984). *Noise-Induced Transitions*, (Springer-Verlag, Berlin).
Inomata A., (1983). Phys. Lett. A **87**, 387.
Inomata A., (1984). Phys. Lett. A **101**, 253.
Izús G., Deza R., Ramírez O., Wio H.S., Zanette D.H., Borzi C. (1995). Phys. Rev. E **52**, 129.
Izús G., Wio H.S., Reyes de Rueda J., Ramírez O., Deza R. (1996). Int. J. Mod. Phys. B **10**, 1273.
Izús G., Deza R. and Wio H.S., (1998). Phys. Rev. E **58**, 93.
Izús G.G., Deza R.R., Wio H.S., (2009). Centre Europ. Phys. J. **71**, 80.
Janssen H.K. (1976). Z. Phys. B **23**, 377.

Jarzynski C. (2008). Eur. Phys. J. B **64**, 331-340.
Jülicher F., Adjari A. and Prost J. (1997). Rev. Mod. Phys. **69**, 1269.
Jumarie G. (2007). Appl. Math. Lett. **20**, 846-852.
Jung P. and Hänggi P. (1987). Phys. Rev. A **35**, 4464.
Jung P. and Hänggi P. (1989). J. Opt. Soc. Am. B **5**, 979.
Kac M., (1949). Trans. Am. Math. Soc. **65**, 1-13.
Kac M., (1959). *Probability and Related Topics in the Physical Sciences* (Interscience, NY).
Khandekar D.C., Lawande S.V., (1975). J. Math. Phys. **16**, 384.
Khandekar D.C., Lawande S.V., (1986). Phys. Rep. **137**, 115-229.
Khandekar D.C., Lawande S.V., Baghwat K.V. (2000). *Path integral methods and their applications* (World Scientific, Singapore).
Kleinert H., (1989a). Mod. Phys. Lett. A **4**, 2329.
Kleinert H., (1989b). Phys. Lett. B **224**, 313.
Kleinert H., (1990a). Phys. Lett. B **236**, 315.
Kleinert H., (1990b). *Path integrals in quantum mechanics, statistics and polymer physics* (World Scientific, Singapore).
Kuperman M. N., Wio H. S., Izús G. and Deza R. (1998). Physical Review E **57**, 5122.
Kwas M., (2005). J. Math Phys. **46**, 103511.
Landau L., Lifshitz E.M., (1958). *Quantum Mechanics* (Pergamon, New York).
Langouche F., Roekaerts D., Tirapegui E.(1982). *Functional Integration and Semiclassical Expansions* (D.Reidel Pub.Co., Dordrecht).
Larsen A., Ravndal F., (1988). Am. J. Phys. **56**, 1129.
Lenzi E.K., Ribeiro H.V., Mukai H., et al. (2010), J. Math. Phys. **51**, 092102.
Levit S., Smilansky U., (1985), in *Proc. Winter College on Fundamental Nuclear Physics*, Eds. Dietrich K., DiToro M., Mang H.J., (World Scientific, Singapore).
Lévy, P. (1953), *Random functions: General theory with special references to Laplacian random functions*, University of California Publications in Statistics, **1**, 331390.
Lindenberg K., West B. J. and Masoliver J. ('1989). in *Noise in Nonlinear Dynamical Systems*, edited by Moss F. and McClintock P. V. E., Vol. 1, (Cambridge University Press, Cambridge).
Lindenberg K., Wio H.S., (2003) *Noise Induced Phenomena. A Sampler* in *Proc. Pan-American Advanced Studies Institute. New Challenges in Statistical Physics*,Lindenberg K., Kenkre V., Eds. (AIP,New York).
Lindner B., Garcia-Ojalvo J., Neiman A., Schimansky-Geier L.. Phys. Rep. **392**, 321-424 (2004).
Luciani J.F., Verga A., (1988). J. Stat. Phys. **50**, 567.
Luzcka J. (1988). J. Phys. A: Math. Gen. **21** 3063.
Luzcka J., Hänggi P. and Gadomski A., (1995). Phys. Rev. A **51**, 2933.
MacKeown P.K., (1985). Am. J. Phys. **53**, 880.
Mandelbrot, B.; van Ness, J.W. (1968), *Fractional Brownian motions, fractional noises and applications*, SIAM Review **10** (4), 422-437.
Mangioni S., Deza R., Wio H.S., Toral R., (1997). Phys. Rev. Lett. **79**, 2389.

Mangioni S., Deza R., Wio H.S., Toral R., (2000). Phys. Rev. E **61**, 223.
Mannheim A.R., (1983). Am. J. Phys. **51**, 328.
Manwani A. (2000). *Information-Theoretic Analysis of Neuronal Communication*, PhD Thesis, CALTECH.
Manwani A., Koch C. (1999). Neural Comp. **11**, 1797.
Marinov (1980). Phys. Rep. **60**, 1.
Martin P.C., Siggia E.D., Rose H.A., (1973). Phys. Rev. A **8**, 423-437.
Mazzucchi S. (2009). *Mathematical Feynman path integrals and their applications* (World Scientific, Singapore).
Mikhailov A.S. (1990). *Foundations of Synergetics I*, (Springer-Verlag, Berlin).
Mikhailov A.S., Loskutov A.Yu. (1992). *Foundations of Synergetics II*, (Springer- Verlag, Berlin).
Morandi G., Galleani D'agliano E., Napoli F., Ratto C.F. (1974). Adv. Phys. **23**, 867.
Moss F., (1992). in *Some Problems in Statistical Physics*, edited by G. Weiss (SIAM, Philadelphia).
Nicolis G. (1995). *Introduction to Nonlinear Science*, (Cambridge U.P., Cambridge).
Nozaki D., Mar D.J., Griegg P., Collins J.D. (1999). Phys. Rev. Lett. **72**, 2125.
K. Oldham and J. Spanier, (1974) *The Fractional Calculus* (Academic Press, New York)
Onsager L., Machlup S., (1953a). Phys. Rev. **91**, 1505.
Onsager L., Machlup S., (1953b). Phys. Rev. **91**, 1512.
Pak N., Sokman I., (1984). Phys. Rev. A **30**, 1629.
Pak N., Sokman I., (1984). Phys. Lett. A **100**, 327.
Papadopoulos G.J., Devreese J.T., Eds., (1978). *Path Integrals and their Applications in Quantum, Statistical and Solid State Physics* (Plenum, New York).
Pelster A., Kleinert H., (1997). Phys. Rev. Lett. **78**, 565.
Pesquera L., Rodriguez M.A., Santos E., (1983). Phys. Lett. A **94**, 287.
Phytian R. (1977). J. Phys. A **10**, 777.
Phytian R., Curtis W.D. (1980). J. Phys. A **13**, 1575.
Podlubny, I. (1998), *Fractional Differential Equations* (Academic Press, New York)
Prato, D. and Tsallis, C. (1999), Phys. Rev. E **60**, 2398.
Reimann P.(2002). Phys. Rep. **361**, 57-265.
Risken H. (1983). *The Fokker-Planck Equation*, (Springer-Verlag, Berlin).
Rodriguez M., Abramson G., Wio H.S. and Bru A. (1993); Phys. Rev. E **48**, 829.
Sagues F.,(1984). Phys. Lett. A **104**, 1.
Sagues F., San Miguel M. and Sancho J.M., (1986). Zeit. f.Physik **55**, 269.
Sagues F., Sancho J.M,, Garcia-Ojalvo J. (2007). Rev. Mod. Phys. **79**, 829-882.
Salem L.D., Wio H.S., (1986). Phys. Lett. A **194**, 168.
Sánchez A.D., Rodriguez M.A. and Wio H.S. (2000); in *Material Instabilities: Proc. 1st. Latin American Summer School*, Eds. Martinez J., Wrner C.H. and Walgraef D. (World Scientific).

Sancho J. M. and San Miguel M., in *Noise in Nonlinear Dynamical Systems*, edited by Moss F. and McClintock P. V. E., Vol. 1, (Cambridge University Press, Cambridge).

Sa-yakanit V., *et al.* Eds, (1989). *Path Integrals from meV to MeV*, (World Scientific, Singapore).

Schulman L.S. (1981). *Techniques and Applications of Path Integration* (Wiley, New York).

Sebastian, K.L. (1995), J. Phys. A: Math. Theor. **28**, 4305.

Sengupta M. (1986). Am. J. Phys. **54**, 1024.

Shankar R., (1980). *Principles of Quantum Mechanics* (Plenum, N.Y.).

Singh N. (2008). J. Stat. Phys. **131**, 405-414.

Strier D., Drazer G., Wio H.S. (2000). Physica A **283**, 255.

Taniguchi T., Cohen E.G.D. (2007). J. Stat. Phys. **126**, 1-41.

Tarasov V.E. and Zaslavsky G.M. (2008); Comm. Nonlin. Sci. and Numer. Simul. **13**, 248.

Tsallis C. (1988). Stat. Phys. **52**, 479.

van den Broeck C., Parrondo J.M.R. and Toral R. (1994), Phys. Rev. Lett. **73**, 3395.

Van den Broeck C. (2010), J. of Stat. Mech.: Theory and Experiment, P10009.

Van den Broeck C., Kumar N. and Lindenberg K. (2012), Phys. Rev. Lett. **108**, 210602.

van Kampen N. (2004). *Stochastic Processes in Physics and Chemistry* 2nd Edition, (Elsevier, The Netherlands).

van Kampen N. (1985). Phys. Rep. **124**, 69.

Walgraef D., (1997). *Spatio-temporal pattern formation*, (Springer-Verlag, New York).

Wang, K.G. and Lung, C.W. (1990), Phys. Lett. A **151** 119.

Watson G.N., (1962). *A Treatise in the Theory of Bessel Functions* (Cambridge Univ. Press., England).

Wiegel F.W., (1986). *Introduction to path integral methods in physics and polymer science* (World Scientific, Singapore).

Wiener N., (1923). J. Math. **2**, 131.

Wiener N., (1924). Proc. London Math. Soc. **22**, 454.

Wiener N., (1930). Acta Math. **55**, 117.

Wiesenfeld K. and Moss F., (1995). Nature **373**, 33.

Wio H.S. (1983). Zeit.fur Phys. **B50**, 357.

Wio H.S. (1986). in *Phase Space Approach to Nuclear Dynamics*, Eds. DiToro M., Norenberg W., Rosina M. and Stringari S. (World Scientific, Singapore).

Wio H.S. (1988). J. Chem. Phys. **88**, 5251.

Wio H.S. (1990). *Introducción a las Integrales de Camino* (Univ.Illes Balears, Palma de Mallorca, Spain).

Wio H.S. (1993). in *Instabilities and Nonequilibrium Structures IV*, Tirapegui E. and Zeller W. Eds. (Kluwer, Dordrecht).

Wio H.S. (1994), *An Introduction to Stochastic Processes and Nonequilibrium Statistical Physics*, (World Scientific, Singapore).

Wio H.S. (1996), Phys. Rev. E **54**, R3075.

Wio H.S. (1997), in *4th. Granada Seminar in Computational Physics*, Eds. P. Garrido and J. Marro (Springer-Verlag, Berlin), pg. 135.

Wio H.S., (1999), *Application of Path Integration to Stochastic Processes. An Introduction*, chapter in. *Fundamentals and Applications of Complex Systems*, Ed. G. Zgrablich, (Nueva Edit. Univ., Univ.Nac. San Luis), pg. 253.

Wio H.S. (2005). *Noise Induced Phenomena and Nonextensivity*, Europhysics News nr.6, 197.

Wio H. S., López A. and Alascio B. (1978). J. Magn.and Magnetic Mat. **8**, 71.

Wio H., Colet P., San Miguel M., Pesquera L., Rodriguez M.A., (1989). Phys. Rev. A **40**, 7312.

Wio H. S., Briozzo C. B. and Budde C. E. (1993a). in *Path Integration. Trieste 1991*, Cerdeira H., Lundqvist S., Mugnai D., Ranfagni A., Sa–yakanit V. and Schulman L.S. Eds.(World Scientific, Singapore)

Wio H. S. and Nicolini F. G. (1993b). in *Path Integration. Trieste 1991*, Cerdeira H., Lundqvist S., Mugnai D., Ranfagni A., Sa–yakanit V. and Schulman L.S. Eds.(World Scientific, Singapore)

Wio H., Zanette D.H., (1993). Phys. Rev. E **47**, 384.

Wio H., Budde C., Briozzo C., Colet P., (1995). Int. J. Mod. Phys. B **9**, 679.

Wio H.S., Castelpoggi F. (1997). in *Unsolved Problems of Noise*, pg. 229, Ch.R. Doering, L.B. Kiss and M. Shlesinger Editors (World Scientific, Singapore).

Wio H. S., Kuperman M. N., Castelpoggi F., Izús G. and Deza R. (1998). Physica A, **257**, 275.

Wio H.S., Von Haeften B., Bouzat S. (2002). in *Proc. 21st STATPHYS*, Physica A **306**, 140-156.

Wio H.S., Deza R.R. (Invited) (2007). Europ.Phys.J-Special Topics **146**, 111.

Wio H.S., Deza R.R. and López, (2012), *An Introduction to Stochastic Processes and Nonequilibrium Statistical Physics*, Revised Edition (World Scientific, Singapore)

Index